Hybrid Anisotropic Materials for Wind Power Turbine Blades

Hybrid Anisotropic Materials for Wind Power Turbine Blades

Yosif Golfman

CRC Press
Taylor & Francis Group
Boca Raton London New York

CRC Press is an imprint of the
Taylor & Francis Group, an **informa** business

CRC Press
Taylor & Francis Group
6000 Broken Sound Parkway NW, Suite 300
Boca Raton, FL 33487-2742

First issued in paperback 2018

© 2012 by Taylor & Francis Group, LLC
CRC Press is an imprint of Taylor & Francis Group, an Informa business

No claim to original U.S. Government works

ISBN-13: 978-1-4398-6858-4 (hbk)
ISBN-13: 978-1-138-38213-8 (pbk)

Library of Congress Cataloging-in-Publication Data

Golfman, Yosif.
    Hybrid anisotropic materials for wind power turbine blades / Yosif Golfman.
        p. cm.
    Includes bibliographical references and index.
    ISBN 978-1-4398-6858-4 (hardback)
    1. Turbines--Blades--Materials. 2. Wind turbines--Materials. 3. Anisotropy. I. Title.

TJ267.5.B5G65 2012
621.4'5--dc23
                                                                2011032559

Visit the Taylor & Francis Web site at
http://www.taylorandfrancis.com

and the CRC Press Web site at
http://www.crcpress.com

Contents

Preface

It is possible that, in the future, the rapidly growing green technologies can replace the nuclear industry.

The U.S. Department of Energy (DOE) wants to improve the cost effectiveness of wind power even further by cutting the price to \$5/MWh (megawatt hour). Efficiency of wind power turbine blades is 15 to 20 times more efficient than solar power panels. As blades get larger and designs more complex, maintaining blade quality becomes more challenging. Automation is needed to increase manufacturing precision and process control. The development of automated and repeatable production techniques including greater use of robotics and process controls for lamination, blade finishing, trimming, grinding, painting, materials handling, pultrusion, and inspection would not only improve quality control, but reduce production costs. Thus, wind turbine design innovations will be necessary across the manufacturing sector.

The DOE (U.S. Department of Energy, 2011) sees advantages to new innovations:

Significant reductions of greenhouse gas (GHG) emissions

Carbon dioxide (CO_2) is the principal GHG in the Earth's atmosphere. Approximately 40 percent of total U.S. CO_2 emissions come from power generation facilities. Since substantial amounts of coal and natural gas fuels would be displaced, the 20 percent wind scenario could reduce CO_2 emissions by the year 2030 by 825 million metric tons—25 percent of the CO_2 emissions from the nation's electric sector in the no-new-wind scenario. This reduction could nearly level projected growth in CO_2 emissions from electricity generation.

Siting strategies and environmental effects

The report examines siting issues and effects that an increase in wind power facilities may have on compatible land uses, water use, aesthetics, and wildlife habitats. However, wind energy avoids many of the undesirable environmental impacts from other forms of electricity production, such as impacts from fuel mining, transport, and waste management.

Unlike fossil fuel and nuclear generation, which use significant quantities of water for power plant cooling, wind power generation consumes no water during operations. Generating 20 percent of U.S. electricity from wind would reduce water consumption in the electric sector by 2030 by 17 percent.

The DOE report examines some of the costs, challenges, and key impacts of generating 20 percent of the nation's electricity from wind energy by 2030.

Specifically, it investigates requirements and outcomes in the areas of technology, manufacturing, transmission and integration, markets, environment, and siting.

The modeling done for this DOD report estimates that wind power installations with capacities of more than 300 gigawatts (GW) would be needed for the 20 percent wind scenario. Increasing U.S. wind power to this level from 11.6 GW in 2006 would require significant changes in transmission, manufacturing, and markets. This report presents an analysis of one specific scenario for reaching the 20 percent level and contrasts it to a scenario of no wind growth beyond the level reached in 2006. Major assumptions in the analysis have been highlighted throughout the document and have been summarized in the appendices. These assumptions may be considered optimistic.

However, the goal remains 20 percent wind energy by the year 2030. Power proportional to square length has required that the increase efficiency length of a blade will be increased from 50.5 m (166 ft.) to 100 m (332 ft.). This scale of blade length requires significant architectural changes because the loading involved is not only flapwise causing lift, but also edgewise, such as a cantilevered beam. These goals need a redesign process, which would include:

1. Robust manufacturing processes
2. Minimizing the optimal number of shear webs (spars) and their placement
3. Development of a solid laminate versus sandwich or stiffened panels for unsupported areas
4. Selected proportion of the fiberglass and carbon fiber around the chord of the blade from leading edge to trailing edge and then down its length
5. Finding the strongest carbon fiber/fiberglass percentage combination and layout stacking sequence that can be achieved at the lowest weight
6. Improving the optimum balance of least-complex, lightest-weight composite tape and fabric layup sequencing with minimal ply drops that results in the lowest fabrication costs within a 24-hour production cycle

The number of the defects (laminates, porosity, and etc.) depends on technological manufacturing methods. Today, TPI Composites, Inc. (Scottsdale, Arizona) selects the patented infusion molding process SCRIMP (Seemans composite resin infusion molding process), and, obviously, if technological processes change to the robust use of the automation fiber placement Ingersoll Machine, the number of defects will be dramatically reduced as well as the cost of fabricated blades. Dynamic fatigue life will be increased and length of life will be predicted more accurately.

Reference

U.S. Department of Energy, 2011, Wind and Water Power Program, http://www1.eere.energy.gov/wind/wind_animation.html, June 28.

Acknowledgments

Dr. Mel Schwartz, an editor for SAMPE (Society for the Advancement of Material and Process Engineering), helped me greatly by editing my articles which have been published in the *SAMPE Journal* and *Journal of Advanced Materials* over a period of 10 years. Thanks very much for this help and for referencing my book.

Dr. Jerome Fanucci, president and CEO of KaZaK Composites Inc., a leader in pultrusion technology, encouraged me to develop pultrusion methods for variable cross-section turbine blades.

Dr. Scott W. Beckwith, technical director of SAMPE, and a specialist in control technology and composites, offered me excellent advice and encouragement when writing this book. I thank you very much for this.

I also would like to thank Dr. A. Brent Strong, editor-in-chief of the *Journal of Advanced Materials,* who sent my articles to reviewers and helped me to update these articles.

To members of Taylor & Francis Group, I want to thank Jonathan W. Plant, senior acquiring editor; Jay Margolis, project editor; Amber Donley, project coordinator; Arlene Kopeloff, editorial assistant; and Scott H. Hayes, prepress supervisor, for technical and structural advice while preparing the manuscript.

Nelson Landrau, president of Neo-Advent Technologies, and Dr. Alex Nevoroskhin, vice president for Department of Defense proposals, showed me a number of new ways to structure this book. I am grateful for their assistance.

About the Author

Yosif Golfman is director of structured materials at Neo-Advent Technologies LLC. He has been involved in the research and development of composites since 1960, starting as a research composite engineer at Shipbuilding Technology Institute in Leningrad.

Golfman had earned his PhD at the institute, investigating the influence of technological factors, fiber prepregs and resins on the strength of fiberglass propellers and blades.

He emigrated to the United States in 1988, working as a research engineer at Ad Tech Systems Research in Dayton, Ohio, as a mechanical engineer at Foster-Miller Inc. in Waltham, Massachusetts, and as a process mechanical engineer at Spectran Inc. in Sturbridge, Massachusetts.

At Spectran Inc., he developed vapor deposition coating control Nan ceramics particles on draw optical glass fibers. At Neo-Advent Technologies Inc. he worked on the design, technology development, and manufacturing of lightweight nanoscale structures parts based on ceramic, thermoplastics, liquid polymers, and carbon fiber textiles for aerospace applications and space vehicles.

He has written more than 30 articles in journals such as the *Journal of Advanced Materials, Reinforced Plastics & Composites*, and *JEC Magazine Composites*. The *Journal of Advanced Materials* published a special edition of Golfman's articles.

He has been a SAMPE member since 1990 and worked for different proposals for composite applications for U.S. Department of Defense SBIR/STIR solicitations.

1

Applications throughout the World

1.1 Introduction

The rapid expansion of wind power is partly the result of increased turbine size and the addition of new wind farms. It is also due to technological improvements in the wind turbines themselves. Better rotor blade aerodynamics, more efficient generators, and improved supervisory control systems make today's wind turbines more efficient than ever before.

This enhanced technology presents engineers with new challenges. A wind turbine is a complex system in which a variety of subsystems must work together as efficiently as possible. The subsystems include mechanical devices, such as rotor blades, gearboxes, hydraulic or electric drives for setting blade pitch angles, electrical yaw drives, generators, and the connection to the electrical grid. All of these are monitored by a complex supervisory control system that must respond in a specific way to varying environmental conditions, such as changing wind speeds.

1.2 Large Wind: Blades and Rotors

There are more than 20,000 wind turbines in operation in Germany generating nearly 24,000 MW, which is roughly 7 percent of the electricity consumed in Germany.[1]

Over the past several years, wind turbines, rotors, and blades have become larger and their designs more complex. For example, designs common in the 1980s involved blades that were about 8 m long, but today they can reach 40 m (or more) for land-based applications and 60 m (or more) for offshore applications. To achieve the 20 percent wind scenario, it is projected that this trend toward larger blade sizes and design complexity will need to continue in order to boost power and energy output, increase capacity factors, increase efficiencies, and lower overall capital costs.

This trend poses challenges in scaling up blade and rotor manufacturing production capacity. For example, the use of larger blades requires lighter-weight materials to increase efficiency and performance and decrease weight, which, in turn, reduces load-carrying requirements for towers and other structural components. However, tower designs may be dictated by aerodynamic loads at the greater heights and sweep areas. Lightweight materials, such as fiberglass and carbon fiber, are in high demand globally, not only for wind energy development, but for a variety of other products and components. While there is a high likelihood that businesses will make the necessary investments to expand production of these and other materials and components for wind energy, a robust supply chain would be critical for the 20 percent wind scenario to be achieved.[2]

1.2.1   Key Blade and Rotor Manufacturing Challenges

Improving quality with increasing blade lengths, design complexity, and use of lighter-weight advanced materials present such challenges as:

- Lack of adequate testing facilities for perfecting advanced blade designs to improve efficiency, performance, and manufacturability, while reducing weight and cost
- Lack of adequate analysis tools for integrating design and manufacturing
- Cost and availability of raw materials for advanced blade designs
- Addressing workforce shortages that affect all aspects of wind energy design, development, manufacturing, and deployment
- Addressing the high costs and logistical problems of transporting blades to construction sites for assembly

One of the technical challenges in mass-producing larger blades with more complex designs and lighter-weight advanced materials is quality control. Faults and defects add to overall costs, can detract substantially from wind turbine performance and efficiency, and can lead to later problems in the field with operation, maintenance, and blade life. The need for better blade quality creates added burdens in ramping up manufacturing capacity for some of the supply chain industries, such as forging, casting, fiberglass, carbon fiber, and bearings. In addition, existing production processes are also highly labor intensive, which adds to costs and increases the difficulties in achieving high levels of manufacturing precision and process control.

1.3   How Wind Turbines Work

A wind turbine will deflect the wind, even before the wind reaches the rotor plane. This means that we will never be able to capture all of the energy in the wind when using a wind turbine. Imagine this image: We have the wind coming from the right and we use a device to capture part of the kinetic energy in the wind. (In this case, we use a three-bladed rotor, but it could be some other mechanical device). The wind energy turns the kinetic wind energy into operational energy, e.g., electricity and/or heat. The converter is based on a rotor driven by the wind, thereby extracting the power of:[3]

$$P = arAv^3 \tag{1.1}$$

where

a is an aerodynamic efficiency constant; (a = .50 - .59)

r the density of air, (at sea level and at 15°C, according to International Standard Atmosphere, air has a density of approximately 1.29 kg/m³, at 200C density 1,204 kg/m³).

A is the area of rotor plane, (A = pD²) and v is the wind velocity (in a storm, it can reach 20–25 m/sec.). If rotor diameter equals 30m, P = 15 MW

Wind is a form of solar energy and is caused by the uneven heating of the atmosphere by the sun, the irregularities of the Earth's surface, and rotation of the planet. Wind flow patterns are modified by the Earth's terrain, bodies of water, and vegetation. Humans use this wind flow, or motion energy, for many purposes: sailing, flying a kite, and even generating electricity.

The terms *wind energy* or *wind power* describe the process by which the wind is used to generate mechanical power or electricity. Wind turbines convert the kinetic energy in the wind into mechanical power. This mechanical power can be used for specific tasks, such as grinding grain or pumping water, or a generator can convert this mechanical power into electricity.

So, how do wind turbines make electricity? Simply stated, a wind turbine works just the opposite of a fan. Instead of using electricity to make wind, like a fan, wind turbines use wind to make electricity. The wind turns the blades, which spin a shaft, which connects to a generator and this generates electricity. To take a look inside a wind turbine to see the various parts, go to www1.eere.energy.gov. View the wind turbine animation to see how a wind turbine works.[4]

This aerial view of a wind power plant (Figure 1.1) shows how a group of wind turbines can make electricity for the utility grid. The electricity is sent through transmission and distribution lines to homes, businesses, schools, and so on. More specific details are described in Chapter 7.

Many wind farms have sprung up in the Midwest in recent years, generating power for utilities. Farmers benefit by receiving land lease payments from wind energy project developers (Figure 1.2).

GE Wind Energy's 3.6 MW wind turbine is one of the largest prototypes ever erected. Larger wind turbines are more efficient and cost effective.

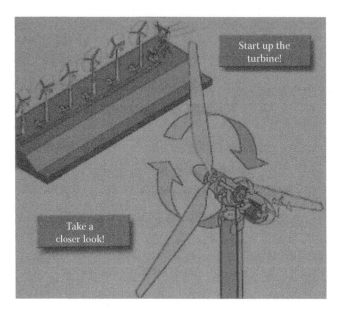

FIGURE 1.1
Aerial view of a wind power plant. (From Wind and Water Power Program, Technologies, 1–2. www1.eere.energy.gov (accessed September 2010).)

FIGURE 1.2
Land lease project for farmers. (From Wind and Water Power Program, Technologies, 1–2. www1.eere.energy.gov (accessed September 2010).)

1.3.1 Types of Wind Turbines

Modern wind turbines fall into two basic groups: the horizontal axis variety, as shown in the photo, and the vertical axis design, like the eggbeater-style Darrieus model, named after its French inventor.

Horizontal axis wind turbines typically have either two or three blades. The three-bladed wind turbines are operated "upwind," with the blades facing into the wind.

1.3.2 Sizes of Wind Turbines

Utility-scale turbines range in size from 100 kW to as large as several megawatts. Larger turbines are grouped together into wind farms that provide bulk power to the electrical grid.

Single, small turbines, below 100 kW, are used for homes, telecommunications dishes, or water pumping. Small turbines are sometimes used in connection with diesel generators, batteries, and photovoltaic systems. These systems are called hybrid wind systems and are typically used in remote, off-grid locations where a connection to the utility grid is not available.

1.3.3 Inside the Wind Turbine

The units that make up a wind turbine (Figure 1.3) include:

Anemometer: Measures the wind speed and transmits wind speed data to the controller.

Blades: Most turbines have either two or three blades. Wind blowing over the blades causes the blades to "lift" and rotate.

Brake: A disc brake, which can be applied mechanically, electrically, or hydraulically to stop the rotor in emergencies.

Controller: Starts up the machine at wind speeds of about 8 to 16 miles per hour (mph) and shuts off the machine at about 55 mph. Turbines do not operate at wind speeds above about 55 mph because they might be damaged by the high winds.

Gearbox: Gears connect the low-speed shaft to the high-speed shaft and increase the rotational speeds from about 30 to 60 rotations per minute (rpm) to about 1,000 to 1,800 rpm, the rotational speed required by most generators to produce electricity. The gearbox is a costly (and heavy) part of the wind turbine and engineers are exploring "direct-drive" generators that operate at lower rotational speeds and don't need gearboxes.

Generator: Usually an off-the-shelf induction generator that produces 60-cycle AC electricity.

High-speed shaft: Drives the generator.

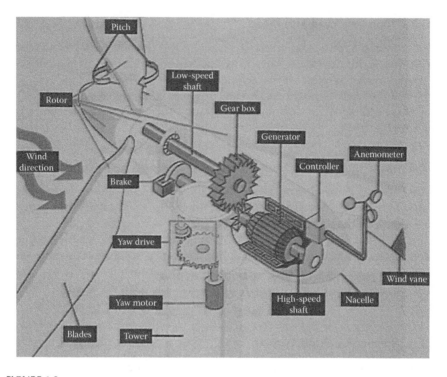

FIGURE 1.3
A view inside the wind turbine. (From Wind and Water Power Program, Technologies, 1–2. www.1eere.energy.gov (accessed September 2010).)

Low-speed shaft: The rotor turns the low-speed shaft at about 30 to 60 rotations per minute.

Nacelle: Sits atop the tower and contains the gearbox, low- and high-speed shafts, generator, controller, and brake. Some nacelles are large enough for a helicopter to land on.

Pitch: Blades are turned, or pitched, out of the wind to control the rotor speed and keep the rotor from turning in winds that are too high or too low to produce electricity.

Rotor: The blades and the hub together.

Tower: Made from tubular steel, concrete, or steel lattice. Because wind speed increases with height, taller towers enable turbines to capture more energy and generate more electricity.

Wind direction: An "upwind" turbine is so-called because it operates facing into the wind. Other turbines are designed to run "downwind," facing away from the wind.

Wind vane: Measures wind direction and communicates with the yaw drive to orient the turbine properly with respect to the wind.

Yaw drive: Upwind turbines face into the wind; the yaw drive is used to keep the rotor facing into the wind as the wind direction changes. Downwind turbines don't require a yaw drive, the wind blows the rotor downwind.

Yaw motor: Powers the yaw drive.

1.3.4   Contradictory Goals

Paul Dvorak, a wind power engineer, mentioned that various control systems with competing goals must interact with each other, and this creates a difficult design challenge.[1] For example, the turbine monitoring system must ensure it generates electricity upon demand. At the same time, it must protect individual parts from unnecessary wear and tear. The supervisory control system must observe both of these contradictory goals.

Normally, the generator is only switched on when the wind reaches a speed of 2 to 4 m/s. Lower wind speeds fail to generate enough power and unnecessarily wear turbine parts. In high winds, controls shut down the generator and set the rotor spin slowly (or not at all) so as to reduce load on the drive train. The system that controls the blades attempts to keep the generator's speed in a relatively narrow range so it can generate the maximum amount of power. Additionally, it also must react to imminent power failures to control or actuate systems within the turbine to prevent the turbine from becoming unstable and destroying itself.

Proper yaw control keeps the turbine facing the wind during normal operation. The yaw controller also must ensure that the nacelle doesn't turn in the same direction all the time so that the cables in the tower do not twist beyond their limit. Again, it is up to the supervisory controller to reconcile these potentially conflicting goals. Adding to the challenge is the fact that the system is nonlinear, including backlash in the gearbox and friction in its large ball bearings.

1.3.5   Smooth and Continuous Development

All subsystems in a turbine can be simulated and tested as part of an integrated system at an early development stage by using a model-based design. Controller hardware can be tested before building prototypes of the full system. Systems that must eventually work together, such as the pitch and yaw actuators, can be simulated together and matched for best performance.

Wind turbine developers who use this design philosophy profit from a smooth and continuous development process. The models and simulations exist in a single environment and are linked directly to the requirements and specifications. In addition, embedded software can be generated directly from the model. Doing so simplifies communication between various teams and enables integration issues to be discovered earlier in the design process.[1]

1.4 Market for Wind Turbine Composites

1.4.1 Introduction

The combined sales of large wind power plants and small turbines for distributed generations are $4 to 5 billion annually worldwide, and growing. The wind turbine manufacturing business has grown from farms to major industrial plans.

1.4.2 Weight and Cost

Power wind turbine blades offer new business opportunities in manufacturing and materials innovation. Worldwide growth in wind generation since 1994 has been 30 percent or higher annually. The cost of energy from large wind power plants has declined to less than $0.05/kWh. By the end of 2000, the global capacity had passed 17,600 MW,[5,6] and the United States alone more than 1,800 MW, and new installation was completed.[7,8]

The weight and cost is the key to making wind energy competitive with other power sources because research programs have significantly improved the efficiency of the rotor and maximized the energy capture of the machine.

The real opportunity today is through using low-cost materials with high-volume production, while ensuring the reliability, strength, and life expectation. The typical weight and cost of the primary turbine components are shown in Table 1.1.[5]

In addition, there are foundations and conventional ground-mounted systems, including transformers, switching, and other power. Turbine subsystem costs are generally evenly split between rotor, nacelle, drive train power systems, and the tower. The rotor is the highest cost item of the machines and, therefore, must be the most reliable. Towers are normally the heaviest component and could benefit from weight reduction, but lightening the rotor or tower top weight has a multiplayer effect throughout the system, including the foundation.

TABLE 1.1

Turbine Component Weight and Cost

Component	% of Machine Weight	% of Machine Cost[9]
Rotor	37177	20–30
Nacelle and machinery, less	25–40	25
Gearbox and drive train	37025	37178
Generator systems	36927	37025
Weight of top of tower	35–50	n/a
Tower	30–65	37188

1.4.3 Technology Evaluation

Light-weight, low-cost materials are especially important in blades and towers for several reasons: the weight of the blades and rotor is multiplied throughout the machine, it reduces loads, and the machine works in a more stable regime.

The tower weight is 60 percent of the weight of the turbine above the foundation because sophisticated light-weight, high-strength materials are often too costly to justify their use. Another technology shift is occurring in the drive train. In some cases, the gearbox is being eliminated by employing variable speed generators and solid state electronic converters that produce utility quality alternating current (AC) power. This trend began in small machines and is now being incorporated in turbine sizes from 100 kW to 3 MW.[9] Other trends in wind turbine technology are discussed in detail in the *Renewable Energy Technology Characterizations* published by the Electric Power Research Institute (EPRI)[10] with DOE support.

1.4.4 Market and Turbine Components Material Data

1.4.4.1 Wind Energy Market Dynamics

New generation wind turbine designs are continuing to push up the power-generating capacity of wind turbines. A common trend among larger capacity designs is the use of longer blades. Covering a larger area effectively increases the tip-speed of a turbine at a given wind velocity, thus increasing the energy extraction capability of the system.

Denmark-based Vestas Wind Systems A/S once again topped the list of wind turbine original equipment manufacturers, according to Dan Shreve, director with MAKE Consulting. The strategic advisory firm, which focuses on the global wind industry, recently released its annual report that tracks global wind turbine manufacturing trends: Wind Turbine OEMS Global Market Share Report for 2009.[11]

In North America, Canada saw significant growth and the United States posted another record year. According to the American Wind Energy Association's *Year End 2009 Market Report*, released in January 2011, 9,922 MW of new capacity was installed, breaking all previous industry records. The additional capacity increased the U.S. wind plant fleet by 39 percent, bringing the country's total wind power generating capacity to more than 35,000 MW from 25,237 MW at the end of 2008.[11]

According to a report published by *China Daily* in February, there has been a dramatic shift in market demand in recent years from 37.5 m (123 ft.) blades, to blades that are 40.3 m (132 ft.).

At the end of 2009, wind power in China accounted for 25.1 gigawatts (GW) of electrical generating capacity and the country has identified wind power as a key growth component of its economy. China is now the largest producer of wind turbines and the second-largest producer of wind power after the United States.[12]

According to the Global Wind Energy Council, China is expected to remain one of the main drivers of global growth in the coming years with annual additions of more than 20 GW by 2014. This development is supported by a very aggressive government policy and the growth of the domestic industry. The Chinese government has an unofficial target of 150 GW of wind capacity by 2020.

1.4.4.2 About Owens Corning

Owens Corning is a leading global producer of glass fiber reinforcements and engineered materials for composite systems and residential and commercial building materials.

The business delivers a broad range of reinforcement products that provide lightweight alternatives to steel, wood, and aluminum, thereby reducing weight and improving energy efficiency. Additional information is available at www.owenscorning.com.

1.4.4.3 Wind Turbine Database

The wind turbine database consists of the power ranges of turbo machines. The wind turbine database was completed from a variety of industrial, DOE laboratory, and existing resources. Much of the wind turbine components' characteristics and weight data came from the DOE, Wind Partnerships for Advanced Technologies program database, their subcontractors, as well as directly from turbine manufacturers, such as TPI Composites, Inc. Twenty-eight types and models of turbines were analyzed,[5] ranging from small models for direct current (DC) battery changing to large grid connected alternating current (AC) machines currently commercially available and being employed in 100 to 200 MW wind power plants. Very large multimegawatt machines being designed for future wind farm applications, both on- and off-shore also were included in the future market. The specific current models, type, and size are listed in Table 1.2.

The actual unit production and sales data incorporated in the market share database is considered proprietary by the manufacturers. These data were used in estimating weights of materials shown in Table 1.3.

Approximately half of the small turbine market measured in MW is direct drive with no gearbox. The trends in design and manufacturing differ between small and large turbines. Small machines tend to use lighter-weight castings in an effort to reduce costs. Many parts in small turbines are die cast aluminum, while in large machines, steel castings or forgings are needed to meet strength and structural fatigue requirements. The size

TABLE 1.2

Power Turbine Rating (kW)

Turbine Manufacturers	Rated Power (kW)
South West Windpower	0.4, 1.0
Bergey	1, 5, 10
Atlantic Oriented Corp.	50
Northern Power Systems	100
Enercon	500, 850
Micon	600, 800
Bonus	600, 1000
Vestas	660, 850, 1650, 2000
Nordex	1000
Mitsubishi	600, 1000

of steel castings for large turbines, especially the blade hub units, is one of the manufacturing challenges.

Material fatigue properties are an important consideration in wind turbine design. During the expected 30-year life of a wind turbine, many of the components will need to be able to endure 4×10^8 fatigue stress cycles. The high cycle fatigue resistance is even more severe than for aircraft, automotive engines, bridges, and most other structures.

1.4.5 Components Development Trends

1.4.5.1 Rotor Blades

Specific development trends of the rotor blades are discussed below.

Most rotor blades are built from fiber/glass/carbon fibers. As the rotor size increases for the large machine, the percentage of carbon fiber increases because high strength and fatigue resistance need to be compensated.

1.4.5.2 Gear Boxes

Most small turbines designed so that batteries can be changed use a variable speed, permanent magnet, and variable frequency generator connected to a rectifier.

As high-power, solid-state electronics are improved, larger machines are likely to use AC-DC-AC cycloconverters. This is based on turbines being developed by northern power systems (100 kW), and in some commercial machines.

This trend will increase the use of magnetic materials in future turbines. Large epicycle gearboxes used in ships may continue to be the drive system for some large turbines.[13]

TABLE 1.3

Percentage of Materials Used in Current Wind Turbine Components

Component Material, % by Weight	Permanent Magnetic Materials	Prestressed Concrete	Steel	Aluminum	Copper	Glass-reinforced Plastic	Wood Epoxy	Carbon-reinforced Plastic
Rotor hub blades			95–100	5		95	(95)	(95)
Nacelle	(17)		(65)–80	3–4	14	1–(2)		
Gearbox Generator	(50)	(20)–65	98–100	(0)–2	(<1)–2 (30)–35			
Frame machinery and shell tower		2	85–(74) 98	9–(50) (2)	4–(12)	3–(5)		

1.4.5.3 Nacelles

The nacelle contains an array of complex machinery including yaw drives, blade pitch change mechanisms, drive brakes, shafts, bearings, oil pumps and coolers, controllers, and more.

1.4.5.4 Towers

Low-cost materials are especially important in towers because towers can represent as much as 65 percent of the weight of the turbine. Concrete in towers has the potential to be low in cost, but may involve nearly as much steel in the reinforcing bars as a conventional steel tower.

References

1. Dvorak, P. 2010. Wind power engineering—Wind power design. Online at: www.windpowerengineering.com (accessed September 22, 2011).
2. TPI Composites, Inc. 2010. 20 percent wind energy by 2030. Online at: www.20percentwind.org (accessed July 20, 2010).
3. Brondsted, P., H. Lilholt, and A. Lystrup. 2005. *Composite material for wind power turbine blades.* Roskilde, Denmark: Annual Reviews, A Nonprofit Scientific Publisher. email: povlbrondsted@risoe.dk, hans.lilholt@risoe.dkaage.Lustrup@risoe.dk
4. DOE. 2010. *Wind and water power program: How wind turbines work.* Washington, D.C.: Department of Energy. www1.eere.energy.gov (accessed July 29, 2010).
5. Ancona, D.F., and J. McVeigh. 2001, *Wind turbine materials and manufacturing fact sheet.* Washington, D.C.: U.S. Department of Energy, Princeton Energy Resources International, LLC.
6. IEA Annual Report. 2001. *Implementing agreement for co-operation in the research and development of wind turbine systems.* Paris: International Energy Agency.
7. AWEA Market Report. 2001. *Wind energy growth was steady in 2000, Outlook for 2001 is bright.* Minneapolis: American Wind Energy Association. Online at: www.awea.org/fag/Global105-2001.PDF (accessed July 29, 2010).
8. American Wind Energy Association, 2001. *Wind project database: Wind energy projects throughout the United States.* Online at: www.awea.org/projects/index.html (accessed July 29, 2010).
9. Department of Energy. 2001. *Wind power today.* Washington, D,C,: DOE/GO-102001-1325.
10. Electric Power Research Institute and DOE. 1997. *Renewable energy technology characterizations.* Valley Forge, PA: EPRI TR-109496.
11. Wind Turbine OEMS Global Market Share Report for 2009, online at: www.windonline.org/2010/03/wtg-oem-market-shares-2009.html (accessed July 29, 2010).

12. Owens Corning. *New structural materials for wind turbine blades.* Toledo, OH: Owens Corning. Online at www.powergenworldwide.com/.../ (accessed July 29, 2010).
13. Legerton, M .L., A. G. Adamantiades, and D. F. Ancona. 1998. Wind power plants. Paper presented at the World Council Conference, Houston, TX. Working Group Paper 3.3,15.

2

Design Wind Power Turbine

2.1 Introduction

Wind technology is measured by the nature of the resource to be harvested. The United States, particularly in the middle of the country from Texas to North Dakota, is rich in wind energy resources, which is now measured at a 50-m elevation. Measuring potential wind energy generation at a 100-m elevation (the projected operating hub height of the next generation of modern turbines) greatly increases the U.S. land area that could be used for wind deployment.

Taking these measurements into account, current U.S. land-based and offshore wind resources are estimated to be sufficient to supply the electrical energy needs of the entire country several times over.

Wind turbines capture the wind's energy with two or three propeller blades that are mounted on a rotor to generate electricity. The turbines sit high atop towers, taking advantage of the stronger and less turbulent wind at 100 ft. (30 m) or more above ground. A blade manufactured from glass/fiber epoxy composites acts much like an airplane wing. When the wind blows, a pocket of low-pressure air forms on the downwind side of the blade. The low-pressure air pocket then pulls the blade toward it, causing the rotor to turn. This effect is called *lift*. The force of the lift is actually much stronger than the wind's force against the front side of the blade, which is called *drag*. The combination of lift and drag causes the rotor to spin like a propeller, and the turning shaft spins a generator to produce electricity. For utility-scale sources of wind energy, a large number of turbines are usually built close together to form a wind farm. Several electricity providers today use wind farm to supply power to their customers. New design and technology concepts for increasing the power of wind farms have been proposed and developed.

For example, without changing the location of the rotor, energy capture can be increased by using longer blades to sweep more area. Blades that are 5 to 16 percent longer can, for the same rated power output, produce a 10 to 35 percent increase in capacity factor. Building these longer blades at an equal or lower cost is a challenge because blade weight must be capped while turbulence-driven loads remain no greater than what the smaller rotor can handle. With the potential of new, structurally efficient airfoils, new materials, passive load attenuation, and active controls, it is estimated that this magnitude of blade growth can be achieved in combination with a modest system cost reduction.

For many reasons, hybrids that are a combination of glass fiber and carbon have more advantages for strength in wind turbine blades.[1]

2.2   New Design Concept

The new design concept for the base of a wind tower that will increase power in one station two to three times is by installation of 30 to 50 machines with propeller blades. Each station of wind blades has its own drive shaft, which connects to a generator. The shaft, using miter and bevel gears, drives the rotation of the blades.

Technology is intended to achieve higher elevations, where the wind resource is much greater, or to access extensive offshore wind resources. Modern wind turbines, which are currently being deployed around the world, have three-bladed rotors with diameters of 70 to 80 m mounted atop 60- to 80-m towers, as illustrated in Figure 2.1. Nacelle enclosures includes: low-speed shaft, gearbox, generator, 1.5 MW, and electrical control.

Typically installed in arrays of 30 to 150 machines, the average turbine installed in the United States in 2006 can produce approximately 1.6 megawatts (MW) of electrical power. Rotating the blades around their long axis to change the angle of attack, with respect to the relative wind as the blades spin around the rotor hub, controls turbine power output. This is called *controlling the blade pitch*. The turbine is pointed into the wind by rotating the nacelle around the tower. This is called *controlling the yaw*. Wind sensors on the nacelle tell the yaw controller where to point the turbine. These wind sensors, along with sensors on the generator and drive train, also tell the blade pitch controller how to regulate the power output and rotor speed to prevent overloading the structural components. Generally, a turbine will start producing power in winds of about 5.36 m/sec. and reach maximum power output at about 12.52 m/sec. to 13.41 m/sec. The turbine will pitch or feather the blades to stop power production and rotation at about 22.35 m/sec. Most utility-scale turbines are upwind machines, meaning that they operate with the blades upwind of the tower to avoid the blockage created by the tower.

The amount of energy in the wind available for extraction by the turbine increases with the cube (the third power) of wind speed. Therefore, a 10-percent increase in wind speed creates a 33-percent increase in available energy. A turbine can capture only a portion of this cubic increase in energy, though, because power above the level for which the electrical system has been designed, referred to as the *rated power*, is allowed to pass through the rotor.

FIGURE 2.1
Modern wind turbines. (From TPI Composites, Inc. *20 percent energy by 2030.* Online at www.20percentwind.org. With permission.)

In general, the speed of the wind increases with the height above the ground, which is why engineers have found ways to increase the height and the size of wind turbines while minimizing the costs of materials. But land-based turbine size is not expected to grow as dramatically in the future as it has in the past. Larger sizes are physically possible; however, the logistical constraints of transporting the components via highways and of obtaining cranes large enough to lift the components present a major economic barrier that is difficult to overcome. Many turbine designers do not expect the rotors of land-based turbines to become much larger than about 100 m in diameter, with corresponding power outputs of about 3 to 5 MW. Rotor Hub Tower, 80 m Minimum Power approximately equal to the length squared ($W \approx L^2$ [L is length of blade]).[1]

2.3 Rotor Design

Typically, a modern turbine will cut in and begin to produce power at a wind speed of about 5 m/sec. It will reach its rated power at about 12 to 14 m/sec., where the pitch control system begins to limit power output and prevent generator and drive train overload. At around 22 to 25 m/sec., the control system pitches the blades to stop rotation, feathering the blades to prevent overloads and damage to the turbine's components.

The job of the rotor is to operate at the absolute highest efficiency possible between cut-in and rated wind speeds, to hold the power transmitted to the drive train at the rated power when the winds go higher, and to stop the machine in extreme winds. Modern utility-scale wind turbines generally extract about 50 percent of the energy in this stream below the rated wind speed compared to the maximum energy that a device could theoretically extract, which is 59 percent of the energy stream (see The Betz Limit).

Most of the rotors on today's large-scale machines have an individual mechanism for pitch control, i.e., the mechanism rotates the blade around its long axis to control the power in high winds. This device is a significant improvement over the first generation of fixed-pitch or collective-pitch linkages because the blades can now be rotated in high winds so they can be feathered out of the wind. This reduces the maximum load on the system when the machine is parked. Pitching the blades out of high winds also reduces operating loads, and the combination of pitchable blades with a variable-speed generator allows the turbine to maintain generation at a constant, rated-power output. The older generation of constant-speed rotors sometimes had instantaneous power spikes up to twice the rated power. Additionally, this pitch system operates as the primary safety system because any one of the three independent actuators is capable of stopping the machine in an emergency.

The Betz Limit

Not all of the energy present in a stream of moving air can be extracted; some air must remain in motion after extraction. Otherwise, no new, more energetic air can enter the device. Building a wall would stop the air at the wall, but the free stream of energetic air would just flow around the wall. On the other end of the spectrum, a device that does not slow the air is not extracting any energy, either. The maximum energy that can be extracted from a fluid stream by a device with the same working area as the stream cross section is 59 percent of the energy in the stream. This maximum is known as the Betz Limit because wind turbine pioneer Albert Betz first derived it.

The No. 1 target for advancement is the means by which the energy is initially captured: the rotor. No indicators currently suggest that rotor design novelties are on their way, but there are considerable incentives to use better materials and innovative controls to build enlarged rotors that sweep a greater area for the same or lower loads. Two approaches are being developed and tested to either reduce load levels or create load-resistant designs. The first approach is to use the blades themselves to attenuate both gravity- and turbulence-driven loads (see section 2.5). The second approach lies in an active control that senses rotor loads and actively suppresses the loads transferred from the rotor to the rest of the turbine structure. These improvements will allow the rotor to grow larger and capture more energy without changing the balance of the system. They also will improve energy capture for a given capacity, thereby increasing the capacity factor.[2] The capacity factor achieved by new wind turbines in 2004 and 2005 reached 36%.

The gearbox is a critical component of gear-driven wind turbines. It converts the motion of the rotating blades to a higher speed, which is required to drive an electric generator. The load and torque characteristics of wind applications are such that the design and manufacture of wind turbine gearboxes is very demanding and requires specialist expertise and highly specialized manufacturing capabilities. Another innovation already being evaluated at a smaller scale by Energy Unlimited, Inc. (EUI: Boise, Idaho) is a variable-diameter rotor that could significantly increase capacity factor. Such a rotor has a large area to capture more energy in low winds and a system to reduce the torque required to drive an electric generator.

2.4 Transmission for Wind Turbine Blades

A transmission and gearbox provide speed and torque conversions from a rotating power source to another device using gear ratios. The term *transmission* refers to the whole drive train, including gearbox, clutch, prop shaft (for rear-wheel drive), differential, and final drive shafts. The most common use is in motor vehicles where the transmission adapts the output of the internal combustion engine to the drive wheels. Such engines need to operate at a relatively high rotational speed, which is inappropriate for starting, stopping, and slower travel. The transmission reduces the higher engine speed to the slower wheel speed, increasing torque in the process. Transmissions also are used on pedal bicycles, fixed machines, and anywhere else rotational speed and torque needs to be adapted. Often, a transmission will have multiple gear ratios (or simply "gears") with the ability to switch between them as speed varies. This switching may be done manually (by the operator) or automatically. Directional (forward and reverse) control may be provided as well. Single-ratio transmissions also exist, which simply change the speed and torque (and sometimes direction) of motor output.

In motor vehicle applications, the transmission will generally be connected to the crankshaft of the engine. The output of the transmission is transmitted via a driveshaft to one or more differentials, which in turn drive the wheels. While a differential also may provide gear reduction, its primary purpose is to change the direction of rotation.

Conventional gear/belt transmissions are not the only mechanism for speed/torque adaptation. Alternative mechanisms include torque converters and power transformation (e.g., diesel-electric transmission, hydraulic drive system, etc.). The line between the wind turbine blade and generator is shown in Figure 2.2.

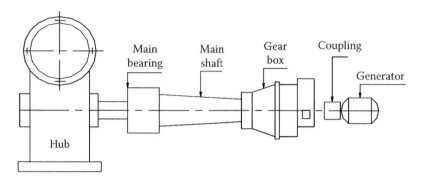

FIGURE 2.2
The line between the wind turbine blade and generator.

Transmissions include main bearing, main shaft, gearbox, coupling, and generator. Most modern gearboxes are used to increase torque while reducing the speed of a prime mover output shaft (e.g., a motor crankshaft). This means that the output shaft of a gearbox will rotate at a slower rate than the input shaft, and this reduction in speed will produce a mechanical advantage causing an increase in torque. A gearbox can be set up to do the opposite and provide an increase in shaft speed with a reduction of torque. Some of the simplest gearboxes merely change the physical direction in which power is transmitted.

Many typical automobile transmissions include the ability to select one of several different gear ratios. In this case, most of the gear ratios (often simply called *gears*) are used to slow down the output speed of the engine and increase torque. However, the highest gears may be "overdrive" types that increase the output speed.

Gearboxes have found use in a wide variety of different—often stationary—applications, such as wind turbines.

Transmissions also are used in agricultural, industrial, construction, mining, and automotive equipment. In addition to ordinary transmission equipped with gears, such equipment makes extensive use of the hydrostatic drive and electrical adjustable-speed drives.

The simplest transmissions, often called gearboxes to reflect their simplicity (although complex systems are also called gearboxes in the vernacular), provide gear reduction (or, more rarely, an increase in speed), sometimes in conjunction with a right-angle change in direction of the shaft (typically in helicopters). These are often used on powered agricultural equipment, since the axial shaft is at odds with the usual need for the driven shaft, which is either vertical (as with rotary mowers) or horizontally extending from one side of the implement to another (as with manure spreaders, flail mowers, and forage wagons). More complex equipment, such as silage choppers and snow blowers, have drives with outputs in more than one direction.

The gearbox in a wind turbine converts the slow, high-torque rotation of the turbine into a much faster rotation of the electrical generator. These are much larger and more complicated than the PTO gearboxes in farm equipment. They weigh several tons and typically contain three stages to achieve an overall gear ratio from 40:1 to over 100:1, depending on the size of the turbine. (For aerodynamic and structural reasons, larger turbines have to turn more slowly, but the generators all have to rotate at similar speeds of several thousand rpm.) The first stage of the gearbox is usually a planetary gear, for compactness and to distribute the enormous torque of the turbine over more teeth of the low-speed shaft. Durability of these gearboxes has been a serious problem for a long time.

2.5 Blades Design

Larger rotors with longer blades sweep a greater area, which increases energy capture. Simply lengthening a blade without changing the fundamental design, however, would make the blade much heavier. In addition, the blade would incur greater structural loads because of its weight and longer moment arm. Using advanced materials with higher strength-to-weight ratios can control blade weight and resultant gravity-induced loads. Because high-performance materials, such as carbon fibers, are more expensive, they would be included in the design only when the payoff is maximized. These innovative airfoil shapes hold the promise of maintaining excellent power performance, but have yet to be demonstrated in full-scale operation (see Figure 2.1).[1] This airfoil shape keeps a basic cutting wind load, bending moments relative to axis *x*, *y*, and twisting moments. Here, C is a center of acting wind forces. We have assumed that the power turbine blade is presented as a curvature plate with variable thickness and force acting on it. Acting forces are the wind load F, rotating forces on angle θ, bending and torsion moments. Pagano and Soni[3] assumed that the blade rotates about the *x* axis, which is parallel to X with a constant angular velocity. In our case, velocity will be changed; it depends on the strongest winds. We have X, Y, Z Cartesian coordinates and Polar coordinates R, θ, X. We see section A-A with acting forces. Fz is an axial force and Fy is a tangential force. Mz and My are bending moments and Mx is a rotating moments relative to axis X.

The area of cross section is denoted by A and the volume of the region above this plane by V. ρ is designated the mass density, ω the angular velocity, r = radius of rotating blade, and the components of the force acting on the cross section A are given by:

$$Fz = \rho\omega^2 r_v \sin\theta \; r^2 x \, dr \, d\theta \, dx \qquad (2.1)$$

$$Fy = \rho\omega^2 r_v \cos\theta \; r^2 x \; dr \; d\theta \; dx \qquad (2.2)$$

While the moments about the x,y,z axis are:

$$Mx = \rho \; z_c\omega^2 r_v \; r^2 \; \sin\theta \; dr \; d\theta \; dx \qquad (2.3)$$

$$My = \rho \; \omega^2 r_v \; x \, r^2 \; \cos\theta \; dr \; d\theta \; dx \qquad (2.4)$$

$$Mz = \rho \; \omega^2 r \, x \, r^2 \; \sin\theta \; dr \; d\theta \; dx \qquad (2.5)$$

Here, z_c is a distance from point C—the center of acting hydrodynamic forces to axis y. θ is the angle of the rotating blade. Equation (2.3) to Equation (2.5) can be solved as:

$$Fz = - \rho\omega^2 x \; r^3/3 \; \cos\theta \qquad (2.6)$$

$$Fy = \rho\omega^2 x \; r^3/3 \; \sin\theta \qquad (2.7)$$

$$M_x^R = - \rho \; z_c \; \omega^2 x \; \cos\theta \; r^3/3 \qquad (2.8)$$

$$M_y^R = - \rho \; \omega^2 x^2 \; \cos\theta \; r^3/6 \qquad (2.9)$$

$$M_z^R = - \rho \; \omega^2 x^2 \; \sin\theta \; r^3/6 \qquad (2.10)$$

Bending and torsion moment will be changed if speed wind blades change (Figure 2.3).

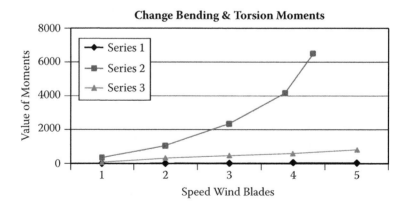

FIGURE 2.3
Caption? (From Bronstead, Lithoit, and Lystrup. 2005. *Composite materials for wind power turbine blades.* Copenhagen: Technical University of Denmark, Riso National Laboratory for Sustainable Energy. With permission.)

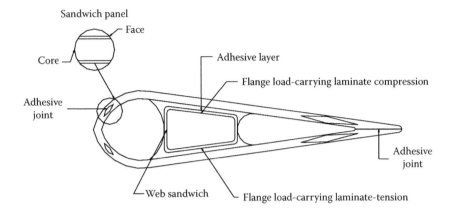

Cross-section principal rotoblade with different construction elements

FIGURE 2.4
Cross-section principal with rotor blade with different blade construction elements. (From TPI Composites, Inc. *20 percent energy by 2030*. Online at www.20percentwind.org. With permission.)

Cross-section principal with rotor blade with different blade construction elements is shown in Figure 2.4. The thickness shell has been carried to the load of the blade and adhesive layers transfer this load to the web beams.[5] The parts design section of the turbine blade can be seen in Figure 2.5, And, the power turbine blade with coordinate system is shown in Figure 2.6.

2.5.1 Theoretical Investigation

In the linear performance relationship between acting forces and linear deformation in matrix form, the following was established:

$$
\begin{array}{cccccc}
F_z, F_y & Q_\varepsilon & Q_\varepsilon x & Q_{\varepsilon R} & Q_{\varepsilon \theta} & \varepsilon \\
M_z^R & Q_{x\varepsilon} & Q_x & Q_{xR} & Q_{x\theta} & \Upsilon x \\
M_y^R & Q_{R\varepsilon} & Q_R x & Q_R & Q_{R\theta} & \Upsilon_R \\
M_x^R & Q_{\theta\varepsilon} & Q_\theta x & Q_\theta y & Q_\theta & \Upsilon_\theta
\end{array}
\tag{2.11}
$$

Here:
R, θ, X   = current polar coordinates;
$\varepsilon, \Upsilon x, \Upsilon_R, \Upsilon_\theta$ = current deformation existing from the acting forces Fz, Fy and bending moments Mz, My, and torsion moment Mx; Qki = coefficient of stiffness in polar coordinates; a Qki = Qik (k,I = ε, x, R, θ).

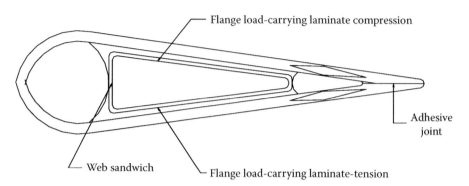

Cross section turbine blade with large core area

FIGURE 2.5
The parts design section of the turbine blade.

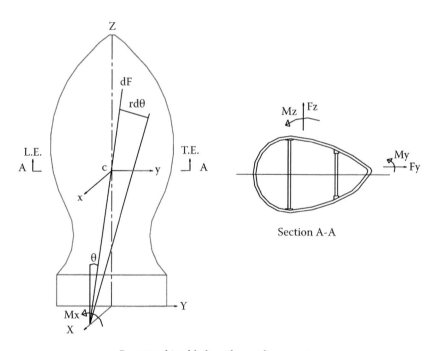

Power turbine blade with coordinate system

FIGURE 2.6
The power turbine blade with coordinate system. (TPI, Inc. 2010. *Manufacturing of utility-scale wind turbine blades*. Online at www.iawind.org/presentation/nolet.pdf. With permission.)

This system follows the theory of Kerhgofa–Klebsha, which is transformed on five independent relations in Equation (2.12).

$$\varepsilon z = \frac{Fz}{E_1 S_1}; \varepsilon y = \frac{Fy}{E_2 S_2}; \Upsilon x = \frac{Mz^R}{E_1 1z}; \Upsilon_R = \frac{My^R}{E_1 1_R}; \Upsilon_\theta = \frac{Mx^R}{GxzT_K} \qquad (2.12)$$

Here, E_1, E_2 = modulus of elasticity; S_1, S_2 = section area; Iz, I_R = moments of inertia for axes x, R; Gxz = shear module; T_x = geometrical stiffness for torsion.

The moment of inertia for axes x, R should be determined as Equation (1.13):

$$Iz = \int_s y^2 dS; Ir = \int_s z^2 dS; \qquad (2.13)$$

The geometrical stiffness for torsion should be determined as:

$$Tx = \int (R^2 \theta dS$$

This force and moment acting alike in gas turbine blades is what was described in Hybrid Anisotropic Materials for Structural Aviation Parts (2010).[4] A modern wind turbine design is shown in Figure 2.7.[1]

FIGURE 2.7
A modern wind turbine design.

The stiffness of the composite is controlled and calculated according to:

$$E_c = \eta \, V_f \, E_f + V_m E_m \tag{2.14}$$

where
  E_c is the stiffness of composite
  η is the orientation factor for the fibers
  V_f is the fiber volume fraction
  E_f is a modulus of elasticity for fiber
  V_m is the matrix volume fraction
  E_m is a modulus of elasticity for matrix

The orientation factor is equal to 1 for aligned parallel fibers loaded along the fiber direction. For the randomly oriented fiber assembly, the orientation factor equals 1/3.[5] The fibers are normally the dominant contributor to the composite properties. In case of wind turbine blade stiffness, calculate as combination of shell stiffness and core stiffness according to:

$$E_c = E_{sh} + E_{cor} \tag{2.15}$$

We reduce the thickness of the shell by increasing stiffness of the core material.

2.5.2   Experimental Investigation

The current coordinate x for the root section we will assume as half of all length. For the middle density of fiberglass, we have 2.5×10^3 kg/m^3. The speed rotation of wind turbine blades varies from 5 to 25 m/sec. We select the angle of twist at 3°. Values of bending and torsion moments are presented in Table 2.1 (see Equation (2.8) to Equation (2.10)).

TABLE 2.1

Values of bending and torsion moments

Bending Moment, $M_x 10^{-6}$, kgm^2/s^2	Bending Moment $M_y 10^{-6}$, kgm^2/s^2	Torsion Moment, kgm^2/s^2
346.4	346.4	13.8
1040.0	1040.0	55.4
2340.0	2340.0	124.7
4160.0	4160.0	221.7
6500.0	6500.0	346.4

2.6   Power Control of Wind Turbines

Wind turbines are designed to produce electrical energy as cheaply as possible. Wind turbines, therefore, are generally designed so that they yield maximum output at wind speeds around 15 m/sec. (30 knots or 33 mph). Its does not pay to design turbines that maximize their output at stronger winds because such strong winds are rare.

In case of stronger winds, it is necessary to waste part of the excess energy of the wind in order to avoid damaging the wind turbine. Thus, all wind turbines are designed with some sort of power control. There are two different ways of doing this safely on modern wind turbines.

2.6.1   Pitch-Controlled Wind Turbines

On a pitch-controlled wind turbine, the turbine's electronic controller checks the power output of the turbine several times per second. When the power output becomes too high, it sends an order to the blade pitch mechanism that immediately pitches (turns) the rotor blades slightly out of the wind. Conversely, the blades are turned back into the wind whenever the wind drops again. Thus, the rotor blades have to be able to turn around their longitudinal axis to select the pitch. During normal operation, the blades will pitch a fraction of a degree at a time, and the rotor will be turning at the same time.[6,7]

Designing a pitch-controlled wind turbine requires some clever engineering to make sure that the rotor blades pitch exactly the amount required. On a pitch-controlled wind turbine, the computer will generally pitch the blades a few degrees every time the wind changes in order to keep the rotor blades at the optimum angle in order to maximize output for all wind speeds.

The pitch mechanism is usually operated using hydraulics or electro-mechanics.

2.6.2   Hydraulic Pitch Control

Hydraulic pitch control for wind turbine blades is possible with rotary unions from Deublin (Waukegan, IL).[8]

Generating steady shaft speed on a wind turbine means constantly adjusting the pitch of each blade to accommodate wind variations. The blades connect to a huge hub mounted to a shaft that turns a gearbox and a generator. Best turbine efficiency calls for a continuous pitch control on the blades.

Blade pitch is powered by either an electric or hydraulic drive. Electric pitch control uses slip rings to conduct power to motors rotating in the hub. Hydraulic systems, on the other hand, use a rotary union to deliver hydraulic power to the drive motor. The industry is split about 45 percent electric and 55 percent for hydraulic controls. The advantage of the hydraulic control is that its power density is higher than electrical equipment and it needs fewer components, making for a simpler system. There are other pluses.

Rotating unions, such as those from Deublin, are precision mechanical devices for transferring fluid from stationary sources to rotating machinery. Ball bearings in a typical union support the rotating components (attached to the machinery) against the stationary component (attached to the fluid supply) and a mechanic seal prevents leaks. While often found on wind turbines, rotating union devices work well in other applications, such as air clutches, gearboxes, machine tool spindles, and more.

2.6.3   Stall-Controlled Wind Turbines

Passive stall controlled wind turbines have the rotor blades bolted onto the hub at a fixed angle. The geometry of the rotor blade profile, however, has been aerodynamically designed to ensure that the moment the wind speed becomes too high, it creates turbulence on the side of the rotor blade, which is not facing the wind. This stall prevents the lifting force of the rotor blade from acting on the rotor.[9]

In Chapter 7, you will see that as the actual wind speed in the area increases, the angle of attack of the rotor blade will increase, until at some point it starts to stall. If you look closely at a rotor blade for a stall-controlled wind turbine, you will notice that the blade is twisted slightly as you move along its longitudinal axis. This is partly done in order to ensure that the rotor blade stalls gradually rather than abruptly when the wind speed reaches its critical value. (Other reasons for twisting the blade are mentioned in the previous section on aerodynamics.)

The basic advantage of stall control is that it avoids moving parts in the rotor itself and a complex control system. On the other hand, stall control represents a very complex aerodynamic design problem, and related design challenges in the structural dynamics of the whole wind turbine, e.g., to avoid stall-induced vibrations. Around two-thirds of the wind turbines currently being installed in the world are stall-controlled machines.

2.6.4   Active Stall-Controlled Wind Turbines

An increasing number of larger wind turbines (1 MW and up) are being developed with an active stall power control mechanism. Technically the active stall machines resemble pitch-controlled machines, since they have pitchable blades. In order to get a reasonably large torque (turning force) at low wind speeds, the machines will usually be programmed to pitch their blades much like a pitch-controlled machine at low wind speeds. (Often they use only a few fixed steps depending upon the wind speed.)

When the machine reaches its rated power, however, you will notice an important difference from the pitch-controlled machines. If the generator is about to be overloaded, the machine will pitch its blades in the opposite direction from what a pitch-controlled machine does. In other words, it will increase the angle of attack of the rotor blades in order to make the blades go into a deeper stall, thus wasting the excess energy in the wind.

One of the advantages of active stall is that one can control the power output more accurately than with passive stall, so as to avoid overshooting the rated power of the machine at the beginning of a gust of wind. Another advantage is that the machine can be run almost exactly at rated power at all high wind speeds. A normal passive stall-controlled wind turbine will usually have a drop in the electrical power output for higher wind speeds, as the rotor blades go into a deeper stall.

The pitch mechanism is usually operated using hydraulics or electric stepper motors. As with pitch control, it is largely an economic question whether it is worthwhile to pay for the added complexity of the machine when the blade pitch mechanism is added.

2.6.5   Individual Pitch Control

The current development trend is toward larger, higher power, wind turbines in order to reduce the ultimate generating cost per kWh. Increasing wind turbine size implies larger rotor diameters and rotor sweep areas. This results in a less uniform wind field over the swept area, which imparts uneven loads on the blades, drive shaft, and turbine structure. These uneven loads increase component wear, reduce efficiency, and increase the amount of maintenance downtime required.

Individual pitch control helps address these issues and provides accurate, real-time load information from each blade.[7]

2.6.6   Other Power Control Methods

Some older wind turbines use ailerons (flaps) to control the power of the rotor, just like aircraft use flaps to alter the geometry of the wings to provide extra lift at takeoff.

Another theoretical possibility is to yaw the rotor partly out of the wind to decrease power. This technique of yaw control is, in practice, used only for tiny wind turbines (1 kW or less), as it subjects the rotor to cyclically varying stress, which may ultimately damage the entire structure.

2.7   Wind Turbine Components

The yaw drive is an important component of the horizontal axis turbine's yaw system. To ensure the wind turbine is producing the maximal amount of electric energy at all times, the yaw drive is used to keep the rotor facing

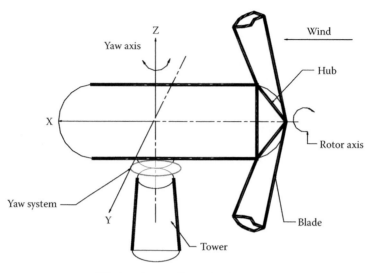

Wind turbine nacelle rotation by yaw system

FIGURE 2.8
Wind turbine nacelle rotation by the yaw system.

into the wind as the wind direction changes. This only applies for wind turbines with a horizontal axis rotor. The wind turbine is said to have a yaw error if the rotor is not aligned to the wind. A yaw error implies that a lower share of the energy in the wind will be running through the rotor area. (The generated energy will be proportional to the cosine of the yaw error) (see Figure 2.8).[10]

2.7.1 History

When the windmills of the eighteenth century included the feature of rotor orientation via the rotation of the nacelle, an actuation mechanism able to provide that turning moment was necessary. Initially, the windmills used ropes or chains extending from the nacelle to the ground in order to allow the rotation of the nacelle by means of human or animal power.

Another historical innovation was the fantail. This device was actually an auxiliary rotor equipped with plurality of blades and located downwind of the main rotor, behind the nacelle in a 90° (approximately) orientation to the main rotor sweep plane. In the event of change in wind direction, the fantail would rotate, thus transmitting its mechanical power through a gearbox (and via a gear-rim-to-pinion mesh) to the tower of the windmill. The effect of the aforementioned transmission was the rotation of the nacelle toward the direction of the wind, where the fantail would not face the wind, thus stop turning (i.e., the nacelle would stop at its new position).[11,12]

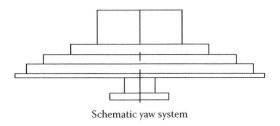

Schematic yaw system

FIGURE 2.9
Schematic yaw system.

The modern yaw drives, even though electronically controlled and equipped with large electric motors and planetary gearboxes, have great similarities to the old windmill concept. They still use a means of mechanical energy "production" (i.e., electric motor), a method to increase the torque (i.e., gearbox), and a gear-rim mounted on the fixed portion of the wind turbine and in constant mesh with the output gear of the said gearbox.

2.7.2 Components

(See the schematic yaw system shown in Figure 2.9.)

2.7.2.1 Gearbox

The gearbox of the yaw drive is a very crucial component since it is required to handle very large moments while requiring the minimal amount of maintenance and also to perform reliably for the whole wind turbine life span (approximately 20 years). Most of the yaw drive gearboxes have input-to-output ratios in the range of 2000:1 in order to produce the enormous turning moments required for the rotation of the wind turbine nacelle.

2.7.2.2 Gear Rim and Pinions

A planetary gearbox reducer is illustrated in Figure 2.10, depicting a schematic representation of a gear rim with teeth on the inner side. The gear rim meshes with four yaw drive pinions.[13]

The gear rim and the pinions of the yaw drives are the components that finally transmit the turning moment from the yaw drives to the tower in order to turn the nacelle of the wind turbine around the tower axis (z axis). The main characteristics of the gear rim are its big diameter (often larger than 2 m) and the orientation of its teeth.

The gear rims with teeth on the outer surface have the advantage of higher reduction ratios in combination with the pinions as well as reduced machining costs over the gear rims with inner teeth. On the other hand, the former configuration requires that the yaw drives be mounted far apart from each other, thus increasing the wind turbine main frame dimensions and costs.

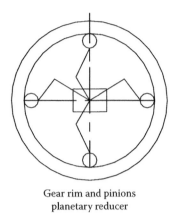

Gear rim and pinions
planetary reducer

FIGURE 2.10
A planetary gearbox reducer.

2.8   Proposal for Robust Redesign Turbine Blades

2.8.1   Introduction

Composite wind turbine blades, helicopter rotors, and fans are manufactured in a laminate form from fiberglass/carbon fibers reinforced epoxy or
vinyl ester resins. The blade skin is made of aramid fiber or Glass C, E, and
Kevlar™ fiber. Additional reinforcement in the leader and trailing edge of
the blade is made for the protection of the artificial loads. For this purpose,
the carbon fiber with reinforced epoxy or vinyl ester resins has been oriented
in various directions.

Changing the fiber direction of the layers can protect wind turbine blades
from the torsion and bending moments. Therefore, the natural frequencies of
wind power blades can be favorably placed in an area outside the operational
rpm (revolution per minute) range.

The purpose of using advanced materials for enlarged rotors is to increase
power, improve structure and aero design, alter active and passive controls,
obtain higher tip speed, and lower acoustic.

2.8.2   Loads Acting Outside Wind Turbine Blades

External loads for wind turbine blades are exposed to wind and gravity. The
speed of the blade is 75 to 85 m/sec, which is much higher than wind speed,
even at storm conditions (25m/sec.). The relative wind directions change
with storm conditions. The blades are exposed to the wind that, through
the lift on the aerodynamic profile, causes loads at right angle to the blades,
which, therefore, react by bending flapwise. The loads are both static, causing
a permanent bending of the blades, and dynamic, causing a fatigue, flapwise

bending due to the natural variations in wind speed. In addition, these static and fatigue load spectra are varied during rotation.

When the blade points upward and downward, respectively, this is caused by the natural wind shear, which is the increase of average wind speed with increasing height over the terrain.

The blades are also exposed to gravity, and this is most pronounced when they are in the horizontal position. These loads cause bending in an angle-wise mode, and a given blade can bend one way on the right-hand side and the opposite way on the left-hand side of the rotor plane. This explains why the edge-wise bending also causes fatigue of the blade material and structure during rotation. The blades are exposed to centrifugal forces during the rotation of the rotor. During the limits of the linear blade velocity, the rotation speed is relatively low, typically from 20 rpm to 10 rpm for large blades. Therefore, the longitudinal tensile loads in the blades are relatively low and normally are not taken into account as a design parameter. The design lifetime of modern wind turbines is normally 20 years, and the corresponding number of rotations is in the order of 10^8 to 10^9.

2.8.3 The Automatic 3-Axial Braiding Process

The automation VARTM processing of large-scale composites structures are presented in *Automated VARTM Processing of Large-Scale Composite Structures* (2004).[14] Instead of using the lay-up of the dry reinforcing fibers in the SCRIMP or VARTM process, it may be more profitable to use automatic 3-axial braiding.

A braider in Figure 2.11 (pos. 3) moves along the mandrel on the rails (pos. 5) and fiber (pos. 2) waving on the mandrel from the rollers (pos. 4) to form a wind blade (pos. 1).

The 3-axial braiding process is finished when we use prepreg and roving fiber. In a case of dry fibers, a smart molding VARTM process is used to open the injection gates. Opening of the injection gates is possible when the area is

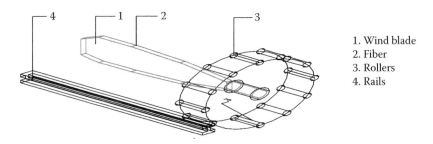

1. Wind blade
2. Fiber
3. Rollers
4. Rails

Braiding wind turbine blade automation process

FIGURE 2.11
Braiding wind turbine blade automation process.

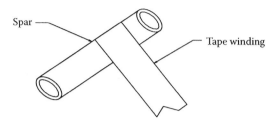

Type winding of a spar for the wind turbine

FIGURE 2.12
Tape winding of a spar for the Nibe wind turbines.

wetted, which ensures complete wet out. Tape winding of a spar for the Nibe wind turbines is shown in Figure 2.12.

The thickness of the spar is varied along the length by varying the length of travel of the tape placement apparatus along the length of the spar. A wedge was manually inserted and withdrawn underneath the tape, at the shear web position of the spar. For each revolution of the mandrel, it is important to control and maintain straight fibers with the desired orientation in the upper and lower part of the spar, which can be accomplished. If you change the direction by 10 degrees, the strength will be immediately reduced by 10 percent.[15]

2.8.4 Pultrusion Process

The pultrusion method for making variable cross-section thermoset articles was patented by M. J. Banassar.[16] J. E. Sumerak developed a method for producing a pultrusion product having a variable cross section using a specially adapted temperature controllable pultrusion die includes:

- Pulling reinforcing fibers that have been impregnated with a heat-curable thermosetting polymeric resin composition through a temperature controllable die
- Heating the temperature-controllable die to a temperature sufficient to effect curing of the thermosetting resin
- Cooling the temperature-controllable die to a temperature that is sufficiently low enough to prevent any significant curing of thermosetting resin passing through the pultrusion die
- Pulling the cured material and a predetermined length of uncured material from the die
- Reshaping the uncured material
- Curing the reshaped material

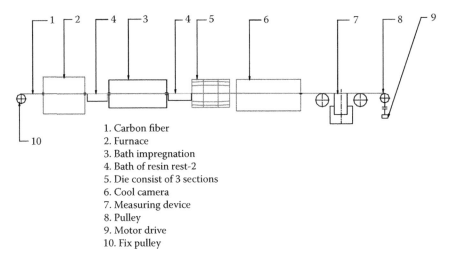

1. Carbon fiber
2. Furnace
3. Bath impregnation
4. Bath of resin rest-2
5. Die consist of 3 sections
6. Cool camera
7. Measuring device
8. Pulley
9. Motor drive
10. Fix pulley

Pultrusion carbon spar for wind turbine blade

FIGURE 2.13
Pultrusion carbon spar for wind turbine blade.

The reshaping step can be used to provide offsets, flanges, bosses, and the like. The method and associated apparatus of the invention provide a relatively simple and inexpensive way of producing fiber-reinforced thermoset plastics having a variable cross section (see Figure 2.12).

Figure 2.13 illustrates the pultrusion carbon spar for wind turbine blades with carbon fiber, pos. 1 going from a fixed pulley, pos. 10, through the furnace, pos. 2, to the bath of resin, pos. 3 and pos. 4. The die, pos. 5, consist of three sections that make the pultrude spars for the two-wind turbine blades. Cooling camera, pos. 6, prevents rest deformations in the inner web spar.

For the design and manufacturing of the prototype YUH-60A UTTAS tail rotor, the pultrusion process[16] was selected. A fully cured, pultruded spar design layout and manufacturing approach was developed. This is important because the pultruded spar consists of the two blades being manufactured simultaneously.

Elastic coupling, which has a significant effect on the dynamic elastic torsion response of the rotor, was discussed in Smith (1994).[17]

Twist coupling effect depends on wind directions, variables, and angle of twist change from ±15° to ±45° (Figure 2.14).

Redesigned pultrusion spars for wind turbine blades are shown in Figure 2.15.

We simultaneously design new spars to be pultruded for blades. These spars are used on root sections of wind turbine blades.

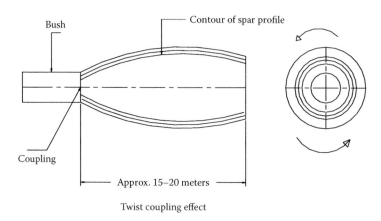

Twist coupling effect

FIGURE 2.14
Twist coupling effect.

FIGURE 2.15
Pultrusion spar for root sections.

The second variant for redesigning shear web profile spars is shown in Figure 2.16. A wood or aluminum mandrel can be used for the tape winding process. Shear web profile has varied dimensions from root sections to middle sections (Figure 2.17).

Other options for making pultrusion web profiles can include the winding shell process with three removable mandrels. It can be seen in Figure 2.18. Carbon fiber strips are indicated in unsupported leading and trailing edges and they add additional stiffness. Shear web pultrusion spars are connected by adhesives in a longitudinal direction, which also resists a bending moment. Torsion moment in a ±45° direction are reinforced by HexForce[a] bias weave fabrics (see Chapter 3, section 3.4).

2.8.5   Shell Curing Mold Prepreg Process

The method of manufacturing sophisticated parts, such as turbine blades by preliminary curing prepreg to 50 percent, was invented in 1969.[18] We used this method to preliminarily seal the rubber film (Figure 2.19).

The purpose is to create a vacuum pressure between the sealed film and the laminate prepreg layers. Therefore, we avoid having to inject liquid resin and this saves time in the curing process. We use power cement forms for molding shell laminates.[19]

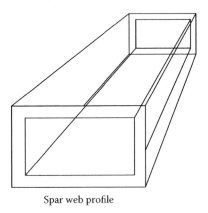

Spar web profile

FIGURE 2.16
Spar web profile.

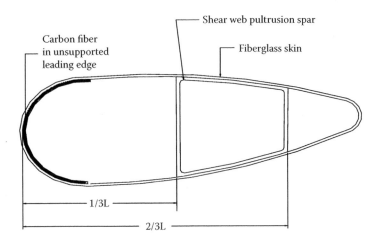

Design shear web pultrusion profile

FIGURE 2.17
Design shear web pultrusion profile.

2.9 Minimizing the Optimal Number of Shear Webs (Spars) and Their Placement

2.9.1 Introduction

It is very important to find a stress function and satisfactory boundary conditions of the shear stress distribution between the outside prepreg layers and the spars in case of a bending moment and a twisting transverse force that act simultaneously.

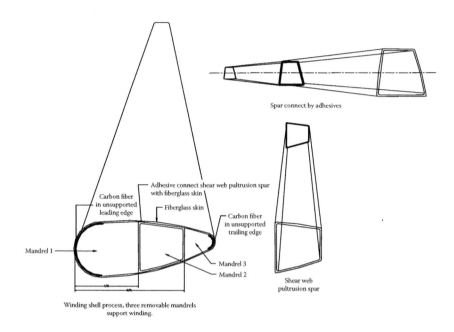

FIGURE 2.18
Winding shell process with three removable mandrels.

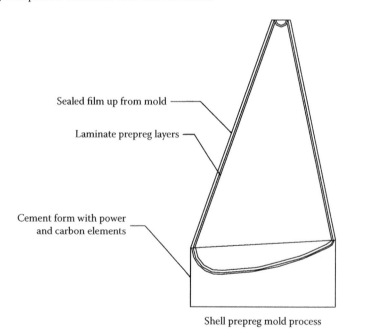

FIGURE 2.19
Shell prepreg mold process.

Designing the shear webs (spars) that carry static and fatigue loads can minimize the wind turbine blade thickness shell.

Shear webs carry the bending and torsion loads, so their placements can be determined by analyzing shear stress distributions.

2.9.2   Shear Web Analysis

The differential equation that determines stress function F for the leading and trailing wind turbine blade is shown as:[4]

$$\frac{1}{G_{yz}^{\ s}+G_{yz}^{\ h}}\frac{\partial^2 F}{\partial x^2} + \frac{1}{G_{xz}^{\ s}+G_{xz}^{\ h}}\frac{\partial^2 F}{\partial y^2} = \frac{2Q(\mu_{zy}^{\ s}+\mu_{zy}^{\ h})y}{(E_z^{\ s}+E_z^{\ h})I} - \frac{Q}{2(G_{xz}^{\ s}+G_{xz}^{\ h})I}\varphi'(Y) + C$$

(2.16)

where,

$E_z^{\ s}$, $E_z^{\ h}$ are the modulus of normal elasticity for skin layers (s) and inner webs (h);

$G_{xz}^{\ s}$, $G_{xz}^{\ h}$ are the modulus of shear stiffness for skin layers (s) and inner webs (h);

I is the moment of inertia for skin and inner webs layers;

F is the stress function that is acting in flexural bending and twisting of the leading and trailing edges of wind turbine blades;

$\mu_{zy}^{\ s}$ and $\mu_{zy}^{\ h}$ are Poisson's ratio for skin layers and inner webs;

Q represents the external dynamic load, which has been acted upon in the center of the point of hydrodynamic stress function;

φ' is the derivative of profile function;

C is a constant of a twist determination.

If the lateral surfaces are free from all external forces, the stress function F for the contour will be equal to zero and all cross sections will satisfy the boundary conditions. From Equation (2.16), the values of the stress function F along the boundary can be calculated for all cross sections.

The shear stresses for the skin plates can be found using Equation (2.17).

$$\tau_{xz}^{\ s} = \frac{\partial F}{\partial y} - \frac{Q}{21_s}\left[x^2 - \varphi(y)\right]$$

(2.17)

The shear stresses for inner webs (spars) can be found using Equation 2.17).
where:

F is a stress function, $\varphi(y)$ is a function of profile;

I_s and I_h are moments of inertia for skin layers and inner web layers.

Now follows:

$$\lambda = \frac{G_{YZ}{}^{S} + G_{YZ}{}^{h}}{E_{XZ}{}^{S} + E_{XZ}{}^{h}}; q = \frac{2(G_{YZ}{}^{S} + G_{YZ}{}^{h})(\mu_{zy}{}^{S} + \mu_{zy}{}^{h})}{E_{z}{}^{S} + E_{z}{}^{h}}; b = \frac{Q}{21} \qquad (2.18)$$

In the future, λ and q will be called the coefficients of anisotropy for a sandwich/hybrid carbon fiber structure. Therefore, Equation (2.16) can be given as:

$$\frac{\partial^2 F}{\partial x^2} + \lambda \frac{\partial^2 F}{\partial y^2} = q \frac{Q}{1} y - \lambda B \varphi'(y) + C \qquad (2.19)$$

Thus, the contour of profile for the leading and trailing edges of wind turbine blades is represented as:

$$f(y) = \lambda B \varphi(y), \text{ to } f'(y) = \lambda B \varphi'(y) \qquad (2.20)$$

Finally, Equation (2.20) can be shown as:

$$\frac{\partial^2 F}{\partial x^2} + \lambda \frac{\partial^2 F}{\partial y^2} = q \frac{Q}{1} y - f'(y) + C \qquad (2.21)$$

Now, we introduce a new stress function, $\phi(u,v)$ in which:

$$u = \lambda/x; v = y, \text{ and } F(x,y) = \phi(x/\lambda, y) \qquad (2.22)$$

After Equation (2.19) is differentiated with respect to x and y, the result is Equation (2.23):

$$\frac{\partial^2 F}{\partial x^2} = \lambda \frac{\partial^2 \phi}{\partial u^2}; \quad \frac{\partial^2 F}{\partial x^2} \frac{\partial^2 \phi}{\partial u^2} \qquad (2.23)$$

In Equation (2.21), stress function F is replaced with a new stress function ϕ using the differential expression in Equation (2.23).

$$\frac{\partial^2 \phi}{\partial u^2} + \lambda \frac{\partial^2 \phi}{\partial v^2} = q \frac{Q}{1} v - f'(v) + C \qquad (2.24)$$

We can designate:

$$\lambda \phi(u,v) = \varphi_1(u,v) \qquad (2.25)$$

As a result, we use a new stress function $\varphi_1(u,v)$ and input into Equation (2.24):

$$\frac{\partial^2 \phi_1}{\partial u^2} + \frac{\partial^2 \phi_1}{\partial v^2} = q\frac{Q}{1}v - f'(v) + C \qquad (2.26)$$

The differential Equation (2.26) had been solved in 1935 for a symmetrical aviation isotropic profile by D. Pinov.[20]

$$\phi1(u,v) = \frac{Q}{81}[u^2 + av^3 + bv^2 + \frac{1}{3a}]* [(b - 1 + 4q)(b + 1) -$$

$$12C_1a]y + C\} [v + \frac{1}{3a} (b - 1 + 4q) \qquad (2.27)$$

In Equation (2.22), a new index stress function u was given by λ/x and y was given by y while in a search for function F, where it was acting in flexure in leading and trailing edges of wind turbine blades and was found to be:

$$F(x,y) = \frac{Q}{81}[\lambda x^2 + av^3 + bv^2 + \frac{1}{3a} \{[(b - 1 + 4q)(b + 1) - 12C_1a]y + C\}$$

$$[v + \frac{1}{3a} (b - 1 + 4q) \qquad (2.28)$$

The first multiplayer is a contour of profile:

$$[\lambda x^2 + av^3 + bv^2 + \frac{1}{3a} [(b - 1 + 4q)(b + 1) - 12C_1a]y + C \qquad (2.29)$$

Here: a, b, C_1, and C are arbitrary coefficients. It is assumed that a = -k and the result is Equation (2.29).

$$[\lambda x^2 - k(y^3 + \gamma_2 y^2 + \gamma_1 y + \gamma_0) = 0 \qquad (2.30)$$

where γ_0, γ_1, γ_2 are designated coefficients of y, and given as:

$$\gamma_2 = \frac{b}{a}; \quad \gamma_1 = \frac{1}{3a} [(b - 1 + 4q)(b + 1) - 12C1a]; \quad \gamma_0 = \frac{C}{a} \qquad (2.31)$$

In Equation (2.30), we replace:

$$y^3 + \gamma_2 y^2 + \gamma_1 y + \gamma_0 = S(y) \qquad (2.32)$$

Reference 20 shows that $S(y)$ can be taken as:

$$S(y) = (y - \alpha)(y - \beta) \qquad (\beta < \alpha) \qquad (2.33)$$

Where the α and β are the distance to edge points of the length profile.

Comparing the coefficients in Equation (2.32) and Equation (2.33) and solving for γ, results in:

$$\gamma_2 = -(2\alpha + \beta);$$

$$\gamma_1 = 2\alpha\beta + \alpha^2; \qquad (2.34)$$

$$\gamma_0 = -\beta\alpha^2$$

Therefore, if coefficients α and β are known, we can determine coefficients γ_2, γ_1, γ_0 using Equation (2.34), and coefficients a, b, C, and C_1 also can be determined. The contour of profile represented in Equation (2.35) is derived from Equation (2.30) as:

$$\lambda x^2 - k(y-\alpha)^2 (y - \beta) = 0 \qquad (2.35)$$

The profile function can be described by using Equation (2.35), which was determined for asymmetrical profiles.

$$f_{1,2}(x)_x = \frac{1}{4} (y_2 - y_1)^2_x$$

where:

$$f_1(x) = a^2 \frac{x}{L_k}(1 + \frac{x}{L_k}) \, x < 0$$

$$\qquad (2.36)$$

$$f_1(x) = a^2 \frac{x}{L_k}(1 - \frac{x}{L_k}) \, x > 0$$

Here, $a = y_2 - y_1$ is a thickness of points; L_k is a width for every cross section. In Equation (2.35), the coefficient of profile k can be shown as:

$$k = \frac{27}{16} \frac{e_k^2 \lambda}{(\alpha - \beta)^2} \qquad (2.37)$$

Here:

e_k is a maximum height of profile;

α is a distance from axis x to exit edge;

β is a distance from axis x to entry edge;

L is a length of profile $(\alpha + \beta)$;

λ is a coefficient of anisotropy.

All the designations are given in Figure 2.20.

The minimum set was determined from Equation (2.35).

$$S'(y) = (y - \alpha)(3y - 2\beta - \alpha) = 0 \qquad (2.38)$$

The abscissa of maximum is a root of Equation (2.38).

Minimizing the optimal number of inner webs for two in each section is possible when we analyze the shear stresses.

Figure 2.18 indicates carbon fiber tape on unsupported leading edge.

$$y_m = \frac{2\beta + \alpha}{3} \qquad (2.39)$$

Obviously, the height of the contour would be one-third the length from the entrance edge (see Figure 2.18).

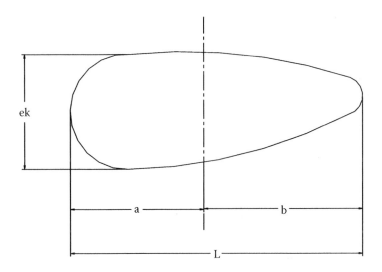

Dimensions of root section of wind turbine blades

FIGURE 2.20

Dimensions of root section of wind turbine blade.

$$e_k = \frac{4}{3^*3^{1/2}} \, (\alpha - \beta) \, [k/\lambda(\alpha - \beta)]^{1/2} \tag{2.40}$$

Therefore, the stress function F includes the equation of aviation symmetrical profiles as expressed in Equation (2.29).

The minimum set was determined from Equation (2.38):

$$S'(y) = (y - \alpha)(3y - 2\beta - \alpha) = 0 \tag{2.38}$$

The abscissa of maximum is a root of Equation (2.38).

Minimizing the optimal number of inner webs for two sections is possible when we analyze the shear stresses. In Figure 2.21, we indicate the carbon fiber type on an unsupported leading edge.

After simplification, the stress function F from Equation (2.29) can be given as:

$$F(x,y) = \frac{Q}{8I\lambda}[\lambda x2 - k(y - \alpha)2(y - \beta)] \, [y + \frac{1}{3a}(b - 1 + 4q)] \tag{2.41}$$

We have determined the derivations of the stress function F:

$$\frac{\partial F}{\partial y} = \frac{Qx}{8I\lambda} \{- 2k(y - \alpha)(y - \beta)[y + \frac{1}{3a}(b - 1 + 4q)] - k(y - \alpha)2 (y + \frac{1}{3a}(b - 1 + 4q)]$$

$$- k \, (y - \alpha)^2(y - \beta) + \lambda x^2) \tag{2.42}$$

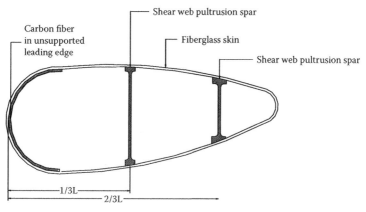

Carbon fiber in unsupported leading edge

Shear web pultrusion spar

Fiberglass skin

Shear web pultrusion spar

1/3L

2/3L

Minimize the optimal number of shear webs

FIGURE 2.21
Minimize the optimal number of shear webs.

The interlaminar shear stresses for the skin turbine blade can be found using Equation (2.42) and Equation (2.47).

$$Y_m = \frac{2\beta + \alpha}{3} \tag{2.43}$$

Obviously the height of the contour would be one-third the length from the entrance edge (see Figure 2.21).

$$e_k = \frac{4}{3^*3^{1/2}}(\alpha - \beta)[k/\lambda((\alpha - \beta)]^{1/2} \tag{2.44}$$

Therefore, the stress function F includes the equation of aviation symmetrical profiles as expressed in Equation (2.29).

After simplification, the stress function F from Equation (2.29) can be given as:

$$F(x,y) = \frac{Q}{81\lambda}[\lambda x^2 - k(y - \alpha^2(y - \beta)[y + \frac{1}{3a}(b - 1 + 4q)] \tag{2.45}$$

We have determined the derivations of the stress function F:

$$\frac{dF}{dy} = -\frac{Q}{4I}(y + \frac{1}{3a}(b - 1 + 4q)$$

$$\frac{dF}{dy} = \frac{Q}{81\lambda}\{- 2k(y - \alpha)(y - \beta)[y + \frac{1}{3a}(b - 1 + 4q)] - k(y - \alpha)^2(y + \frac{1}{3a}(b - 1 + 4q)] - k$$

$$(y - \alpha)^2(y - \beta) + \lambda x^2) \tag{2.46}$$

$$\tau_{xz}' = \frac{Q}{81\lambda}\{- 2k(y - \alpha)(y - \beta)[y + \frac{1}{3a}(b - 1 + 4q)] - k(y - \alpha)^2(y + \frac{1}{3a}(b - 1 + 4q)] - k$$

$$(y - \alpha)^2(y - \beta) + \lambda x^2)$$

$$\tau_{xz}' = -\frac{Q}{4I}(y + \frac{1}{3a}(b - 1 + 4q) \tag{2.47}$$

The interlaminar shear stresses for inner web $\tau_{xz}{}^h$ can be found using Equation (2.47) and Equation (2.48).

$$\tau_{xz}{}^h = \frac{Q}{81\lambda}\{-2k(y-\alpha)(y-\beta)[y+\frac{1}{3a}(b-1+4q)]-k$$

$$(y-\beta = \lambda x^2) - \frac{Q}{21}(y-\alpha)^2(y-\beta); \tag{2.48}$$

$$\tau_{xz}{}' = -\frac{Qx}{4I}(y+\frac{1}{3a}(b-1+4q)$$

2.9.3 Conclusions

1. We calculate the maximum and minimum shear stresses in critical points on the boundary between skin layers and inner web.

   We also have determined the minimum shear stresses in a boundary between skin layers and inner web (spars).

   The interlaminar shear stresses determined in the leading/trailing turbine blade are within the shape of the symmetrical aviation profiles.

   We recommend installing spars (inner webs) at the place x equal 1/3 and 2/3 as shown in Figure 2.17 and Figure 2.21.

2. The traditional method of distribution of shear stresses with the maximum on the neutral axis and minimum on the contour profile does not work according to the formula by Djuravsky.

3. The interlaminar shear stresses inside the field on the neutral axis were equal to zero and matched them to the contour of the shape of the profile.

4. For stress, the function was found satisfactory in the boundary conditions and the shear stress distribution between the outside prepreg layers and the spars in the case of a simultaneously acting bending moment and a twisting transverse force.

2.10   Skin Stiffness and Thickness Blade Calculation

2.10.1   Introduction

The fibers and the matrix are combined into wind turbine blades. The principal design consists of selected shells and core materials, which carry proportional loads that are created by bending and torsion moments.

The leading and trailing edges that are unsupported by inner webs include a carbon fiber, which stiffness is two times more than glass fibers.

2.10.2 Stiffness Calculation

The stiffness of the composite is controlled and calculated according to:[5]

$$E_c = h \, V_f \, E_f + V_m E_m \qquad (2.49)$$

where

E_c is the stiffness of composite
h is the orientation factor for the fibers
V_f is the fiber volume fraction
E_f is a modulus of elasticity for fiber
V_m is the matrix volume fraction
E_m is a modulus of elasticity for matrix

The orientation factor is equal to 1 for aligned parallel fibers loaded along the fiber direction. For the randomly oriented fiber assembly, the orientation factor equals one-third.[5] The fibers are normally the dominant contributor to the composite properties. In case of wind turbine blade, stiffness calculated as combination shell stiffness and core stiffness:

$$E_c = E_{sh} + E_{cor} \qquad (2.50)$$

We reduce thickness of the shell by increasing the stiffness of core material.

2.10.3 Skin Thickness Calculation of Blades

Skin thickness of power turbine blades can be predicted if we know the stresses acting under bending moments Mz, My, and Mx are rotating—torsion moment relative to axis x.

Varying the angle of twist (+15° +45°/-15° -45°), blades create some problems. However, relations between normal and shear stresses can be selected as boundary conditions and strength criteria can be designed as polynomial forth order, such as:

$$a_{ikem} \sigma_{ik} \sigma_{ik} - \left(\frac{(\sigma_{ik} \sigma_{ik})^2 + \sigma_{ik} \sigma_{ik}}{2} \right)^{1/2} = 0 \qquad (2.51)$$

where:

$\delta_{ik} = 0$ if i¹k

This criterion can be used separately for tensile and compressive loads.[21]

Expanding Equation (2.51), criterion of strength for triaxial stress conditions is obtained as:

$$\frac{\sigma^2_x + c\sigma^2_y + b\sigma^2_z + d\tau^2_{xy} + p\tau^2_{yz} + r\tau^2_{zx} + s\sigma_x\sigma_y + t\sigma_y\sigma_x + f\sigma_z\sigma_x}{(\sigma^2_x + \sigma^2_y + \sigma^2_z + \tau^2_{xy} + \tau^2_{yz} + \tau^2_{zx} + \sigma_x\sigma_y + \sigma_y\sigma_x + \sigma_z\sigma_x)^{1/2}} \leq \left[\sigma_{bx}\right]$$

(2.52)

where:

$$c = \frac{X}{Y}; \quad b = \frac{X}{Z}; \quad d = \frac{X}{S_{12}}; \quad p = \frac{X}{S_{23}}; \quad r = \frac{X}{S_{13}};$$

$$s = \frac{4X}{S^{45}_{12}} - c - d - 1; \quad t = \frac{4X}{S^{45}_{23}} - c - b - p; \quad f = \frac{4X}{S^{45}_{13}} - b - r - 1$$

(2.53)

All coefficients get by test laminates spaces.
Here:
X,Y,Z is a tensile (compression) strength in x, y, z directions;
X^{45}, Y^{45}, Z^{45} are the normal strength to act in diagonal directions;
S_{12}, S_{13}, S_{23} are the shear strength to act in plane xy, and out of plane xz, yz;
S^{45}_{12}, S^{45}_{13}, S^{45}_{23} are the shear strength to act in diagonal directions under
　　　　angle 45° in plane xy and interlaminar planes xz, yz;
σ_x, σ_y, σ_z, τ_{xz}, τ_{yz}, τ_{xy}, and τ_{zx} are normal and shear variable stresses, respectively, and depends on loading history.

For biaxial stress conditions, practical calculation strength criteria look like:

$$\frac{\sigma^2_x + c\sigma^2_y + d\tau^2_{xy} + s\sigma_x\sigma_y}{(\sigma^2_x + \sigma^2_y + \tau^2_x + \sigma_x\sigma_y)^{1/2}} \leq \left[\sigma_{bx}\right]$$

(2.54)

The strength of fiberglass comes from textile that has been prepreged by epoxy resins, which are listed in Table 2.2 (all characteristics are in MPa/psi).
Parameters for related strength of fiberglass are shown in Table 2.3.
Equation (2.52) reduces to tensile(compression) stress σ_x, if σ_x equal σ_y and shear stress equal ¼ σ_x and parameters for calculus from Table 2. 2 are used.
In case of tensile load, Equation (2.52) looks like:

$$3.93\sigma_x \leq [\sigma_{bx}]$$

(2.55)

In case of a compression load, Equation (2.52) looks like:

$$5.93\sigma_x \leq [\sigma_{bx}]$$

(2.56)

TABLE 2.2

Strength of Fiberglass

Load Direction	X	Y	Z	X^{45}	Y^{45}	Z^{45}
Tensile	69/1025	46/683,6	2,3/34,18	23/341,8	4,6/68,4	4,6/68,4
Compression	50,6/752	34,27/509,2	64,6/960,4	19,2/285,4	22,31/331,5	22,31/331,5
Shear	S_{12}^{45} 15/222,17	S_{23}^{45} 15/222,17	S_{13}^{45} 15/222,17			

TABLE 2.3

Parameters for Related Strength of Fiberglass

Load Direction	c	b	d	p	r	s	t	f
Tensile	1.50	30.0	4.60	4.60	4.60	11.36	7.76	−17.14
Compression	1.47	.785	3.38	3.38	3.38	7.68	7.48	8.365

Following Goldenblat and Kopnov,[22] the thickness of the cross section thin tube-h depends on torsion moment-T.

$$h = \frac{T}{2A\tau_{xy}} \tag{2.57}$$

where

A is the area enclosed by the center line of tube,

t_{xy} is the shear stress of any sections.

Twist (torsion) moment change from the wind direction and reinforced fiber is very important (Figure 2.22).

2.10.3.1 Experimental Results

We selected boundary conditions for wind turbine blade; the thickness of the shell blade takes place in biaxial stress conditions s_x and s_y, following Goldenblat and Kopnov:[22]

$$\frac{\sigma_x + \sigma_y}{p} = \frac{R_s}{h} \tag{2.58}$$

Here:

σ_x, σ_y = stresses acting in longitudinal and axel directions;

p   = hydrostatic wind pressure,

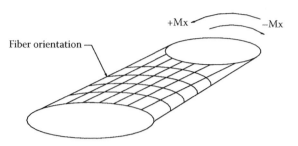

Twist bending moment change from wind directions

FIGURE 2.22
Twist bending moment change from wind directions.

R_c　　= radius of sphere section of root turbine blade,
h　　= middle thickness of skin root section of turbine blade structure.

So, we replace $\sigma_x = pR_s/2h$ and input in Equation (2.58) and get a thickness in a tension zone: $h = 1.965\ p\ R_s$; in compression zone: $h = 2.965\ p\ R_s$. This means, in case of atmospheric pressure $p = 1\text{kg/cm}^2$ and radius of sphere is approximately 1 cm, thickness of shell varies 2 to 3 mm.

2.11　Deflection of Wind Hybrid Blades

The maximum deflection of cantilever bar under aerodynamic force P, which was investigated in Ober, Jones, and Horton:[23]

$$\omega = \frac{PL^3}{8EI} \tag{2.59}$$

Here:
　P　= uniform load;
　L　= length of bar;
　E　= modulus of elasticity;
　I　= moment of inertia.

Engineering approaches for deflection cantilever bar also had been done in work.[24]

With maximum deflection of hybrid cantilever bar with fiberglass covers and carbon inner webs, Equation (2.59) will transform in Equation (2.60).

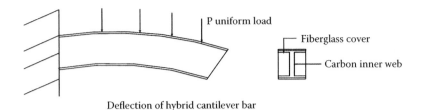

Deflection of hybrid cantilever bar

FIGURE 2.23
Deflection on hybrid cantilever bar.

$$\omega = \frac{PL^3}{8(E_c I_c + E_{iw} I_{iw})} \tag{2.60}$$

Here:

E_c and E_{iw} = modulus of elasticity fiberglass covers and carbon inner webs;

I_{FC} and I_{CW} = moment of inertia fiberglass covers and carbon inner webs.

Deflection on hybrid cantilever bar is shown in Figure 2.23.

The precise solution of cantilever bar deflection with elliptic section are found in the work of Lekhnitski.[25]

$$w = \frac{p_x a^4}{64D'} \left(1 - \frac{x^2}{a^2} - \frac{y^2}{b^2}\right) \tag{2.61}$$

Here:

$$D' = \frac{1}{8}[3D_{11} + 2(D_{12} + 2D_{66})C^2 + 3D_{22}C^4]; \quad C = \frac{a}{b} \tag{2.62}$$

Here:

a is a width section;

b is a height section (Figure 2.24).

The stiffness parameters of the blade structure can be determined as:

$$D_1 = \frac{E^s_{11} h_1^3 + E^h_{11} h_2^3}{12[1 - (\mu^s_{12}\mu^s_{21} + \mu^h_{12}\mu^h_{21})]}; \quad D_2 = \frac{E^s_{11} h_1^3 + E^h_{11} h_2^3}{12[1 - (\mu^s_{12}\mu^s_{21} + \mu^h_{12}\mu^h_{21})]}$$

$$D_{12} = \frac{G^s_{12} h_1^3 + G^h_{12} h_2^3}{12[1 - (\mu^s_{12}\mu^s_{21} + \mu^h_{12}\mu^h_{21})]}; \quad D_3 = D_1(\mu^s_{21} + \mu^h_{21}) + 2D_{12} \tag{2.63}$$

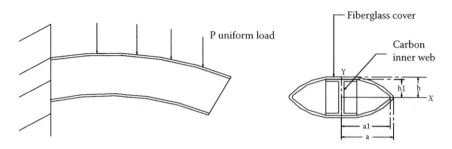

Deflection of hybrid cantilever bar with elliptic section

FIGURE 2.24
Deflection of hybrid cantilever bar with elliptic section.

Here:

E^s_{11}, E^s_{22}, G^s_{12}, μ^s_{12}, μ^s_{21} are the modulus of normal and shear elasticity and Poisson's of ratio for skin layers in axial and transverse directions;

E^h_{11}, E^h_{22}, G^h_{12}, μ^h_{12}, μ^h_{21} are the modulus of normal and shear elasticity and Poisson's of ratio for inner webs in axial and transverse directions.

h_1, h_2 are the thickness of skin and inner web layers.

Moment of inertia fiberglass cover relative axis x is:

$$I_{FC} = 1/4\pi \, (ah^3 - a_1 h_1^3) \tag{2.64}$$

Moment of inertia fiberglass cover relative axis y is:

$$I_{FC} = 1/4\pi \, (ha^3 - h_1 a_1^3)$$

Moment of inertia carbon inner webs relative axis x is:

$$I_{CW} = a_1 h_1^3/12$$

Moment of inertia carbon inner webs relative axis y is:

$$I_{CW} = h_1 a_1^3/12$$

Stress function F (Equation (2.3.26)) is a maximum deflection of wind turbine blade (F = ω):[4]

$$F(x,y) = \frac{Q}{8I\lambda} [\lambda x^2 - k(y-\alpha)2(y-\beta)] \, [y + \frac{1}{3a}(b-1+4q)]$$

Here:

Q represents the external dynamic load, which has been acted upon in the center of the point of hydrodynamic stress function;

x, y are current profile coordinates.

Now follows:
k is a profile coefficient (Equation (2.37))

$$k = \frac{27}{16} \frac{e_k^2 \lambda}{(\alpha - \beta)^2}$$

Here:
e_k is a maximum height of profile;
α is a distance from axis x to exit edge;
β is a distance from axis x to entry edge;
L is a length of profile $(\alpha + \beta)$;
λ is a coefficient of anisotropy.

$$\lambda = \frac{G_{YZ}^S + G_{YZ}^h}{E_{XZ}^S + E_{XZ}^h}; \quad q = \frac{2\,(G_{YZ}^S + G_{YZ}^h)(\mu_{zy}^s + \mu_{zy}^h)}{E_z^s + E_z^h}; \quad b = \frac{Q}{2I}$$

2.11.1   Experimental Investigation

We selected parameters: $\alpha = 2/3$; $\beta = 1/3$; $e_k = y = 1$; $x = 2/3$; coefficients of anisotropy λ means relations transverse stiffness to longitudinal stiffness equals 0.2; $q = .010$; $b = 1.5$; external dynamic load of atmospheric pressure equals 1 kg/m². Profile coefficient k is equal to .189. Deflection is calculated equal to 4.08 m. Static test shows deflection of huge fiberglass blades equal 6 m.

2.11.2   Conclusion

1. Deflection of huge length turbine blades can be determined by having the structures constructed of skin fiberglass and carbon inner webs.

2. Stiffness of carbon inner webs can be increased by reducing deflection of hybrid turbine blades.

References

1. NASA. 20 percent wind energy by 2030. Online at: www.1.eere.energy.gov/windandhydro/pdfs/41869pdf
2. Ashwill. 2004. Windpower-Wikipedia, the free encyclopedia, en.wikipedia.org/wiki/wind_power.
3. Pagano, N. J. and S .R. Soni. 1988. Strength analysis of composite turbine blades. *Journal of Reinforced Plastics and Composites* 7(6): 558–581.

4. Golfman, Y. 2010. *Hybrid anisotropic materials for structural aviation parts*. Boca Raton, FL: Taylor & Francis Group.
5. Brondsted, P., H. Lilholt, and A. Lystrup, 2005, Composite material for wind power turbine blades. Roskilde, Denmark: Annual Reviews, A Nonprofit Scientific Publisher. email: povl.brondsted@risoe.dk, hans.lilholt@risoe.dk, aage.Lustrup@risoe.dk
6. Danish Wind Industry Association (updated June 1, 2003). Online at: http://www.windpower.org/en/tour/wtrb/powerreg.htm
7. www.moog.com/markets/.../wind-turbines/blade-pitch-control/
8. Dvorak, P. 2009. Hydraulic pitch control for wind-turbine blades. Online at: www.windpowerengineering.com/design/mechanical/gearboxes/hydraulic-pitch-control-for-wind-turbine-blades/
9. Ragheb, M. 2009, Control of wind turbines, May 6. https://netfiles.uiuc.edu/mragheb/www/NPRE%20475%20Wind%20Power%20Systems/Control%20of%20Wind%20Turbines.pdf
10. Wikimedia Foundation, Inc., GNU Free Documentation License, Version 1.2 or any later version published by the Free Software Foundation; with no Invariant Sections, no Front-Cover Texts, and no Back-Cover Texts.
11. Gasch, R. and J. Twele. 2004. *Wind power plants*. Berlin: Solarpraxis.
12. Burton, T. et al. 2001. *Wind energy handbook*. New York: John Wiley & Sons.
13. The wind turbine. 2010. www.generalplastics.com/wind turbine
14. D. Heider and J. W. Gillespie. 2004. Automated VARTM processing of large-scale composite structures. *Journal of Advanced Materials* 36 (4).
15. Sumerak, J. E., and B. N. Solon. 1996. Pultrusion method for making variable cross-section thermoset articles. US Patent 5,556,496, filed September 17,1996.
16. Bonassar, M. J. 1980. MM&T fiber-reinforced plastic helicopter tail rotor assembly (pultruded spar). Ft. Rucker, AL: U.S. Army Aviation R&D Command Final Report (for August 1975–October 1978).
17. Smith, E. C.1994, Vibration and flutter of stiff-in plain elasticity tailored composite rotor blade mathematical and computed modeling. *Rotorcraft Modeling* (Special edition) 20 (1–2).
18. Golfman, Y. 1969, Method manufacturing sophisticated parts like turbine blades by preliminary curing prepreg to 50 percent. US Patent 263860.
19. Golfman, Y., D. M. Gek, B. E. Bachareva, and N. P. Sedorov. 1968. Power cement forms for manufacturing screw blades using hot pressing method. *Shipbuilding Technology* 5.
20. Golfman, Y. 1991. Strength criteria for anisotropic materials. *Journal of Reinforced Plastics and Composites* 10(6): 542–556.
21. Chou, P. C. C., and N. J. Pagano. 1967. *Elasticity, tensor, dyadic and engineering approaches*. New York: Van Nostrand Co., 141-143.
22. Goldenblat, I. I. and V. A. Kopnov. 1968. *Resistance of fiberglass*. Moscow: Mashinostroenie.
23. Ober, L., F. D. Jones, and H. L. Horton. 1990. *Machinery handbook*. 23rd ed. New York: Industrial Press Inc., 264–265.
24. Pinov, D. 1935, *Solving problem of bending polynomial stress function*. Moscow. Central Institute of Aerodynamics, 209.
25. Lekhnitski, S. G.1968. *Anisotropic plates*. Translated from the second Russian edition by S. W. Tsai and T. Cheron. New York: Gordon & Breach.

3

Materials for Turbine Power Blades, Reinforcements, and Resins

3.1 Materials Requirements

Material requirements concentrate on stiffness, density, and long time fatigue:

- High material stiffness supports aerodynamic performance
- Low-density reduces gravity forces
- Long fatigue life reduces material degradation

Design process focuses on the materials and the influence that technology has on the stiffness and strength of the materials. Development weight–length correlation is illustrated in Figure 3.1.[1]

Symbols indicate different manufacturers and processing technologies. Shown in the figure, the blade length of 55 m weighs approximately 20 tons. The lower end of the curve represents the relatively short blades of lengths from 12 to 15 m. These were common in the early years of wind turbine design (1980s).

The materials property requirements, which include high stiffness, low weight, and long-fatigue life, can be used to select materials and to make a consideration on what deformation exists within a field of stress. The mechanical design of a rotor blade corresponds nominally to a beam, and the metric index is, for this case:

$$M_b = E^{1/2}/r \tag{3.1}$$

where:
E is material stiffness;
r is the material density.

In Figure 3.2, the diagram shows stiffness versus density for all materials. The metric index M_b is equal to 0,003 (lower line) and 0,006 (upper line). The criterion of absolute stiffness, E = 15GPa, is indicated by the horizontal line.

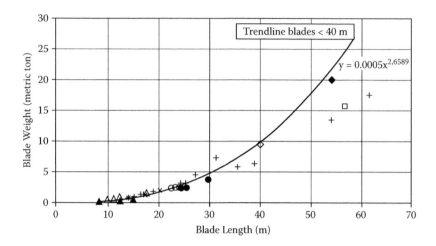

FIGURE 3.1
Development weight length correlation. The triangle-blade length is 10 to 20 m. The circle-blade length is 30 m. The reqular polygon with 4 sides-blade length is 54 m. (From Bronstead, Litholt, and Lystrup. 2005. *Composite materials for wind power turbine blades.* Technical University of Denmark, p. 509. With permission.)

The two lines shown in Figure 3.2 are arbitrary and illustrate lines of materials that are equally good in terms of stiffness and density for a cantilever beam. The lower of the two lines indicates that potential candidate materials are wood, composites, porous ceramics, metals, and ceramics.

The bottom line has a metric index of $M_b = 0.003$ with unit of E in GPa and r in kg/m³. If the metric index is doubled to $M_b = 0.006$, the upper line is valid. A sensible deflection requires a material stiffness of 10 to 20 GPa. To illustrate in Figure 3.2, E = 15 Gpa.

3.2   Structural Composite Material

Structural composite material consists of reinforcements and resins. Variations in reinforcement materials and resin can be used during the winding process, such as E-glass, S-glass, Kevlar™, and carbon fibers.

Table 3.1 lists the mechanical properties of carbon and glass fibers impregnated by epoxy and liquid polymer resins.

Commercial glasses based on silica are obtained by fusing a mixture of materials at temperatures ranging from 1300 to 1600°C.[1] Silica networks (SiO_2) are modified by oxides, such as CaO, BaO, Na_2O, K_2O, and braked Si-O-Si bridges. In doing so, this reduces the temperature in glass-founding furnaces. We analyzed glass for reduced cost, high specific strength, and

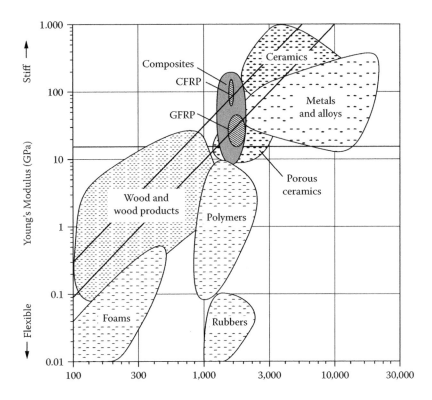

FIGURE 3.2
Diagram correlation between stiffness and density. (From Bronstead, Litholt, and Lystrup. 2005. Composite materials for wind power turbine blades. Technical University of Denmark, p. 510. With permission.)

moderate stiffness. Carbon is used in root blades and leading edges in spite of the high cost because it has specific strength and stiffness.

Fiber and composites properties are shown in Table 3.2.[1]

Glass fiber has an amorphous structure with isotropic properties, but carbon fiber has a crystallographic lattice with a hexagon shape, which is called graphite.

So, the stiffness of a carbon fiber is five times greater than glass fiber, if geometrical axis of a blade consists of fiber orientation (orientation 0°).

Thermoset matrices most used are polyesters, vinyl esters, and epoxy, which are soft and flexible (low stiffness of less than 4 GPa) to help bind the fibers together, work like an interface, and give functional composites a structural purpose. Finally, stiffness of the carbon composites is 4.6 times greater than glass composites and tensile strength is 1.3 times higher, but the density of carbon composites is less. In turbine blades, glass fibers and carbon fibers are based on hybrid construction. The hybrid concept is often a compromise between the improved performance of carbon fibers and their high cost.

TABLE 3.1

Mechanical Properties of Carbon Fiber and Glass Fiber Impregnated by Epoxy and Liquid Polymer Resins

Designation psi (MPA)	Carbon Fiber IM7 + Epoxy	1162U	43741	1166N
Tensile strength	440,000 (3,036)	235,000 (1,62)	254,000 (1,75)	254,000 (1,75)
Tensile modules	25.5×10^6 (1,76 GPA)	15.6×10^6	6.8×10^6 (0.47 GPA)	16.8×10^6 (1,16 GPA)
Compression strength		136,000 (0.938 GPA)	161,000 (1,11 GPA)	76,900 (0,536 GPA)
Compression modulus		14.7×10^6	6.6×10^6 (0.46 GPA)	7.4×10^6 (0,51 GPA)
Flexural strength	240,000 (1,655)	227,000 (1,565)	223,000 (1,54)	230,000 (1,586)
Flexural modulus	21.5×10^6 (1,48 GPA)	16.2×10^6	6.5×10^6 (0,45 GPA)	19.1×10^6 (1,31 GPA)
Short beam shear strength	18,500 (1,28)	12,500 (0,86)	12,500 (0,86)	12,500 (0,86)

Resins, such as epoxy or vinyl/polyester, have demonstrated a capacity for long life. Thermoplastic liquid resins, such as polyphenylene sulfide (PPS) and polyetheretherketone (PEEK), also are used for inner web-cap spars.

The basic commercial parameters for selecting resins material properties are their viscosity and curing time. A practical study of material response is first investigating the material's viscosity.[2]

Viscosity is the measurement of a fluid's resistance to flow. For a liquid under shear, the rate of deformation (or shear rate) is proportional to the shearing stress. Newton's law states that the ratio of the stress to the shear rate is a constant. That constant is viscosity.[3] The common units for viscosity are poise (P), grams per centimeter second g/(cm.s), or dyne seconds per square centimeter (dyn.s)/cm². The centipoise (cP), one hundredth of a poise, is the most common unit. The SI units of viscosity is Pascal seconds (Pa.s and mPa.s)(1 mPa.s = 1 cP).[4,5]

The viscosity of most materials decreases as temperature increases.

The Andrade equation relates the viscosity to temperature as follows:

$$h = A.10B/T \tag{3.2}$$

$$\log \eta = \log A + B/T \tag{3.3}$$

where:
A and B are constants characteristic of the polymer or other material;
T is the absolute temperature.

The viscosity of a polymer at a given temperature can be estimated by knowing the viscosity of two other temperatures. This knowledge allows

TABLE 3.2

Properties of Fibers and Composites

Fiber Type	Stiffness, E_t, GPA	Tensile Strength, MPa	Density, g/cm³	Volume Fraction V_f	Orientation θ	Composite Stiffness, GPA	Tensile Strength, MPa	Density g/cm³
Glass-E	72	3500	2.54	0.5	0°	38	1800	1.87
				0.3	Random	9.3	420	1.60
Carbon	350	4000	1.77	0.5	0°	176	2050	1.49
				0.3	Random	37	470	1.37
Aramid	120	3600	1.45	0.5	0°	61	1850	1.33
				0.3	Random	14.1	430	1.27
Polyethylene	117	2600	0.97	0.5	0°	60	1350	1.09
				0.3	Random	13.8	330	1.13
Cellulose	80	1000	1.50	0.5	0°	41	550	1.35
				0.3	Random	10.1	170	1.29

calculation of the constants A and B and subsequent determination of viscosities at other temperatures. The Andrade equation holds for many polymer solutions and polymer melts well above its glass transition temperatures.[6]

Viscosity is an important parameter for the online process for making vinyl ester/carbon fiber composites. Vinyl ester resins need to have a low viscosity in order to flow into the fibers and to permit good wetting of the carbon fibers. Styrene is used as a diluent to decrease the viscosity of the vinyl ester resins. Thus, the styrene concentration is an important factor that controls the viscosity of vinyl ester resins. Concurrently, viscosity is highly sensitive to changes in temperature, and the molecular weight of the vinyl ester oligomers also affects viscosity significantly. All of these factors need to be studied and understood so that the viscosity of the vinyl ester resins can be controlled.

The effect of temperature and the effect of styrene on viscosity of vinyl ester resins have been reported in literature.[7,8]

Low-viscosity epoxy resins, such as Bisphenol A mPa.s @ 25°C, offer a superior blend of low molecular weight epoxy resin and reactive diluents, designed to produce an easy-flowing, reliable system with low drip and no blush, as well as *no* volatile organic compounds (VOCs). MAS low viscosity epoxy resin is typically used at room temperatures using your choice of hardeners (fast, medium, or slow) in a 2:1 mix ratio, resin:hardener.

Low viscosity epoxy resin can be used in a huge variety of applications including boat hulls and turbine blade structures, which need to completely exclude moisture, as well as the laminating and bonding of nearly all materials.

The viscosity of the epoxy resins must be low to provide efficient flow of the resin into the fiber preforms in an online process. The viscosity of vinyl ester resins decreases with increases in styrene concentration, but styrene concentration can be bad for your health. Vinyl ester resins with varying styrene concentrations provide a system with a wide range of viscosities, which are suitable for different applications. Vinyl ester resins with higher styrene concentrations offer improved resistance to acids and alkalis while those with lower styrene concentrations afford better resistance to solvents.[9] PPG Industries, Shelby, North Carolina, creates fiberglass materials Hybon® 2002 and Hybon® 2026. Hybon is a continuous filament, single-end fiberglass roving, with unique chemistries to work in a variety of resin systems.

Hybon 2002 roving is the industries leading product for wind blade composites. With more than 20 years of wind blade experience, this product has the proven performance for wind blade applications. Hybon 2002 is offered as a single-end roving product in multiple high mechanical properties and is fully compatible with polyester, epoxy, and vinyl ester resin systems.

Hybon 2026 is PPG's highest performance, single-end roving for wind blades. Its unique surface chemistry has an affinity for resin and drives its excellent performance in wind-blade fabrication minimizing dry spots and reworkings. Use of Hybon 2026 wind-blade design achieves higher mechanical performance, most notably in tensile strength and fatigue life, which, in turn, allows one to optimize the composite structure.

TABLE 3.3

Tensile Properties of Roving Materials

Name of Roving Materials	Strength Level	Strength MPa (Kpsi) 600 TEX—2400 TEX
Hybon® 2002	Average Standard dev. N	2290(332)—2230(323) 210(30)—145(21) 37—33
Hybon® 2026	Average Standard dev. N	2790(405)—2660(386) 130(19)—135(20) 13—13

Both materials can be used as reinforced dry laminates in wind turbine blades.[10]

Tensile properties of roving materials are listed in Table 3.3.

An improvement of 20 percent in tensile strength for Hybon 2026 has been very useful in reducing thickness and weight. Van Wazer et al.[8] demonstrated significant reduction of over 50 percent in the level of voids in vinyl ester resin.

Carbon fibers usually are used in root section edges of power turbine blades and inner shear web spars in root sections.

3.3   Resins Advantages: Low Viscosity and Low Curing Time

Epoxies cure best at normal room temperatures, but will cure faster if warmed (30–40°C). Cure times will lengthen as the temperature drops and some epoxies do have a minimum cure temperature, as shown in Table 3.4.[12]

Araldite® 2005 has a minimum curing time and gap filling as well as a simple mixing ratio (2:1 by volume). The curing range is between 20 to 100°C. This provides negligible shrinkage, high shear and peel strength, and heat resistant to 80°C. Araldite 2005 is resistant to water and to a wide range of chemicals.[13,14] Carbon/cyanate ester composites were proposed as replacements for the carbon/epoxy composites with improved out-gassing, microcracking, and moisture absorption characteristics.[15]

To improve service and reduce lead times, Amber Composites (Nottingham, United Kingdom) has increased its unidirectional (UD) prepreg manufacturing capabilities with the addition of a dedicated UD manufacturing line. Amber's resin formulations are now available in a wide range of woven, multiaxial, and unidirectional formats. The new unidirectional systems are designed to complement Amber's range of woven systems, and are available in a wide range of fibers and resin systems.[16]

TABLE 3.4

Cure Time of Adhesive Systems

Adhesive System	Adhesive Type	Curing Time	Estimate Bond Strength, psi
Araldite[a] 2005	Epoxy	20 min, 212°F/100°C	362
Dexter Hysol EA9346	Epoxy	1 h, 250°F/121°C	–
Giba Geigy, Epibond 1545	Epoxy, paste	2 h, 150°F/121°C	–
Araldite 2007	Epoxy	2 h, 248°C/120°C	287
Giba Geigy, XMH 8680	Polyurethane paste	2 h, 120°F/48.8°C	–
Super epoxy	Epoxy	68°F/20°C	281
Loctite 414-N Primer	Cyanoacrylate	68°F/20°C	–

3.4 Rapid Curing Resin System

To assist recreational and industrial markets in their quest for increasingly more efficient composite solutions that permit production rates to increase without affecting investment and labor costs, Hexcel (Stamford, Connecticut) has developed a new epoxy resin system able to break the psychological barrier of the five-minute cure cycle.[17] Curing in only two minutes at 150°C/300°F, HexPly® M77 brings significant advantages to the market, especially as this reactive system also offers a long shelf life of six weeks at 23°C.

Other features, such as a Tg (temperature of transition glass) of 120°C/250°F and good adhesive properties (e.g., when bonding to aluminum), make HexPly M77 highly suitable for large volume production as in the manufacture of skis and automotive components. Being conscious to deliver performance without expense to the environment, HexPly M77 is fully compliant with all regulations.

Based on the HexPly M77 resin system, Hexcel is now launching a range of prepregs based on glass or carbon reinforcements, in multiaxial form as well as woven fabrics.

Cycom® 5320-1 (Cytec Engineered Materials, Anaheim, California) is a toughened, epoxy resin, prepreg system designed for autoclaving, with low temperature curing capabilities and only vacuum bag curing is required. This epoxy has low porosity, low VOCs, and can be recommended for huge turbine blade manufacturing.[18,19] Table 3.5 shows tensile properties IM fiber unidirectional tape: T40/800B impregnated by Cycom 5320-1 (Nominal FAW 145 gsm; nominal resin content 33 percent by weight).

Table 3.5 shows the IM fiber unidirectional tape: T40/800B (Nominal FAW 145 gsm; nominal resin content 33 percent by weight) tensile properties.

TABLE 3.5

Tensile Properties IM Fiber Unidirectional Tape: T40/800B Impregnated by Cycom 5320-1

IM Fiber Unidirectional	Test Method	Condition	Results
0° Tension Strength, ksi (MPa) Modulus Msi (GPa)	ASTM D3039 [0]8	-100°F (-73°C) DRY	372 (2565) 22.6 (156) 0.258
RTA			392 (2703) 22.7 (156) 0.344
90° Tension Strength, ksi (MPa) Modulus Msi (GPa)	ASTM D3039 [90]16	-100°F (-73°C) DRY	12.1 (83) 1.6 (11.0)
RTA			11.7 (81) 1.4 (9.7)
90°/0° Tension Strength, ksi (MPa) Modulus Msi (GPa)	ASTM D3039 [90,0]4S	RTA	190 (1310) 11.9 (82.0)
Unnotched Tension Strength, ksi (MPa) Modulus Msi (GPa)	ASTM D3039 [45,90,-45,0]3S	-100°F (-73°C) DRY	124 (857) 8.5 (58.4)
RTA	135 (931) 8.4 (57.6)		
Filled Hole Tension Strength, ksi (MPa) Modulus, Msi (GPa)	ASTM D6742 [45,0,- 45,0,0,45,90,- 45,0, 0, 45,0,-45,0]S	-100°F (-73°C) DRY	99.1 (683) 14.1 (97.2

TABLE 3.6

Compression Properties

IM Fiber	Test Method Lay Up	Condition	Results
0° Compression Strength, ksi (MPa) Modulus Msi (GPa)	SACMA SRM01R94 [0]8	-100°F (-73°C) DRY	270 (1862) 20.3 (140)
RTA		220°F (104°C)	252 (1737) 20.8 (143)
90° Compression Strength, ksi (MPa) Modulus Msi/ (GPa)	SACMA SRM01R94 [90]16	-100°F (-73°C) DRY	56.9 (392) 1.7 (11.5)
90°/0° Compression Strength, ksi (MPa) Modulus Msi (GPa)	SACMA SRM01R94 [90, 0]4S	RTA	167 (1151) 10.9 (75.1)
Residual strength, ksi (MPa)	ASTM D7137 or SACMA SRM02R94 [45,90,-45,0]4S	RTA	25.6 (176)
Open Hole Compression Strength, ksi (MPa) Modulus, Msi (GPa)	ASTM D6484 [45, 0, -45, 0, 0, 45, 90, -45, 0, 0, 45, 0, -45, 0]S	220°F (104°C) WET3	56.0 (386) 13.8 (95.2)
Filled Hole Compression Strength, ksi (MPa)	ASTM D6742 [45,9 0, -45,0]3S	RTA	79.7 (550)

Table 3.7 represents °the shear properties IM fiber unidirectional tape: T40/800B (Nominal FAW 145 gsm; nominal resin content 33 percent by weight).

Structural Component Manufacture using 8020 Rapi-Ply Technology: www.ambercomposites.com. Issue Ref: TDS/Page 3 of 7 RP8020/05 (July 2009). Rapi-Ply is available in a wide range of high performance reinforcing fabrics, standard configurations including:

Description Rapi-Ply Construction Width Molded thickness

Carbon fabrics

8020 RP 101 HS Carbon 300 g/m² 2/2 twill 6K 43%/1.0 m 0.76 mm

HS Carbon 300 g/m² 2/2 twill 6K 43%

TABLE 3.7

Shear Properties IM Fiber Unidirectional Tape: T40/800B

IM Fiber	Test Method Lay Up	Condition	Results
Short Beam Shear Strength, ksi (MPa)	ASTM D2344 [0]16	-100°F (-73°C) DRY	23.8 (164)
RTA			19.2 (132)
RTA		220°F (104°C) WET3	11.3 (78)
Short Beam Shear Strength, ksi (MPa)	ASTM D2344 [0]30		17.3 (119)
RTA		220°F (104°C) WET3	11.6 (80)
In-Plane Shear Strength, ksi (MPa)	ASTM D3518 [+45, -45]2S	RTA	15.9 (110) 8.4 (58)
Strength at 0.2% offset, 1500-5500 mod pt, ksi (MPa)			7.9 (54) 0.76 (5.2) 0.81 (5.6)
Strength at 0.2% offset, 500-3000 mod pt., ksi (MPa)			
Modulus 1500 - 5500 µin/in, Msi (GPa)			
Modulus 500 - 3000 µin/in, Msi (GPa)			
90° Flexure Strength, ksi (MPa)	ASTM D790 [90]16	RTA	17.5 (121)

8020 RP 102 HS Carbon 300 g/m^2 2/2 twill 6K 43%/1.0 m 0.78 mm

HS Carbon 285 g/m^2 2/2 twill 12K 43%

8020 RP 103 HS Carbon 300 g/m^2 2/2 twill 6K 43%/1.0 m 1.14 mm

HS Carbon 650 g/m^2 2/2 twill 12K 43%

8020 RP 104 HS Carbon 600 g/m^2 2/2 twill 12K 40%/1.0 m 1.45 mm

HS Carbon 650 g/m^2 2/2 twill 12K 40%

Glass fabrics

8020 RP 201 E Glass 390 g/m^2 2/2 twill 35%/1.0 m 0.7 mm

E Glass 390 g/m^2 2/2 twill 35%

8020 RP 202 E Glass 300 g/m^2 8HS 35%/1.0 m 0.55 mm

E Glass 300 g/m^2 8HS 35%

8020 RP 203 E Glass 600 g/m^2 PW WR 35%/1.0 m or 1.2 m 1.1 mm

E Glass 600 g/m^2 PW WR 35%

8020 RP 204 E Glass 450 g/m^2 ±45° 35%/1.27 m 0.88 mm

E Glass 450 g/m2 ±45° 35%

8020 RP 205 E Glass 600 g/m^2 8HS 35%/1.20 m 1.02 mm

E Glass 600 g/m² 8HS 35%

8020 RP 206 E Glass 280 g/m² PW WR 35%/1.00 m 0.48 mm

E Glass 280 g/m² PW WR 35%

All materials are supplied on a roll length of 201 m and supplied with a light tack for use on vertical surfaces with the exception of RP202 and RP205, which are supplied tack-free. The standard resin system is pigmented black. 8020 RAPI.[17]

3.5　Reinforced Material: Carbon Fiber and Glass Fiber Fabrics

Hexcel has developed a patented process to manufacture continuous rolls of carbon fiber fabric in which the warp and weft yarns are oriented on the bias at +/– 45 degrees. These HexForce® bias weave fabrics provide many cost- and weight-saving advantages to customers, notably by reducing prepreg waste and labor operations.[16]

HexForce bias weave fabrics are ideal for high aspect ratio parts that need bias reinforcement, such as aircraft ribs, stringers, spars, and helicopter blades. They also are used in engine nacelles, aircraft barrel sections, and secondary structures. The fabrics are suitable for prepreg and processing and available with HS and IM carbon fibers.

3.5.1　Carbon Fibers

The parameters of a plain weave carbon fiber include: 5.7 oz./sq. yd., 50 in. wide, .012 in. thick, 12.5×12.5 plain weave. This plain weave carbon fiber is the most commonly used type for lightweight aerodynamic parts. Only three layers are typically necessary to produce nonstructural pieces. It wets out quickly and handles easily. Typical full roll length is 100 yards. A 3K, plain weave carbon fiber fabric is shown in Figure 3.3.[19]

Mechanical properties of the IM7 carbon fiber impregnated by epoxy resin are listed in Table 3.8.[12]

Table 3.9 lists the modulus of elasticity and coefficient of thermal expansion for the prepreg laminates based on the IM7 carbon fibers.

The modulus of elasticity and coefficient of thermal expansion for the prepreg laminates based on the K63712 carbon fibers are shown in Table 3.10.

3.5.2　Twill Weave Kevlar®

Twill weave Kevlar is preimpregnated with an epoxy resin system. This material can be stored, shipped, and handled at room temperatures and is

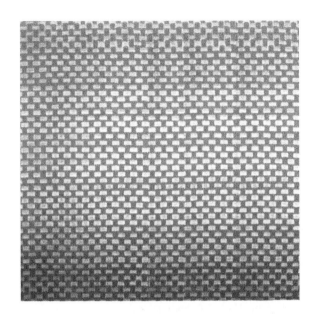

FIGURE 3.3
A 3K, plain weave carbon fiber fabric. (From Carbon Fiber Fabrics: www.composites/vision; www.shopmanic.com/carbon. With permission.)

TABLE 3.8

Mechanical Properties of the IM7 Carbon Fibers

Mechanical Property	Value
Tensile Strength	2,760 MPa
Tensile Modulus	168 GPa
Compression Strength	1,655 MPa
Compression Modulus	148 GPa
Short-beam shear strength	100 MPa
Fiber density	1,770 kg/m^3
Fiber volume fraction	62%

cured using a ramp-up schedule requiring, at most, 310°F. Resin content is 36 to 42 percent. Prepreg 2×2 twill weave Kevlar is pictured in Figure 3.4.

3.5.3 S2-Glass

Parameters of S2-glass: 9 oz./sq. yd., 50 in. wide, .009 in. thick, 57×57 construction, 8H satin weave. This style 6781 S2-glass is identical to the style 7781 E-glass (#543) except it is woven with superior S-2 fibers resulting in extraordinary strength with a superior look and finish. (Typical full roll length is 500 yards.)

TABLE 3.9

Modulus of Elasticity and Coefficient of Thermal Expansion for the Prepreg Laminates Based on the IM7 Carbon Fibers

Material	IM7/PEEK	IM7/PPS	IM7/Boron/ PEEK	IM7/Boron/ PPS
Modulus of Elasticity, E, GPa/Msi	165.7/24.2	165.7/24.2	188.3/27.5	188.3/27.5
Coefficient of Thermal Expansion, CTE, PPMF	0.02	0.06	0.57	0.57

[a] IM7 -60, Matrix -40 percent volume fractions.
[b] 10 percent boron fibers volume fraction.

TABLE 3.10

Modulus of Elasticity and Coefficient of Thermal Expansion for the Prepreg Laminates Based on the K63712 Carbon Fibers

Material	K63712/PEEK[a]	K63712/PPS[a]	K63712/Boron/ PEEK[b]	K63712/Boron/ PPS[c]
Modulus of Elasticity E, GPa/Msi	383.4/56.0	383.4/56.0	384.7/56.2	384.7/56.2
Coefficient of Thermal Expansion CTE, PPM/F	-0.50	-0.48	0.00	0.00

[a] K63712 -60, Matrix -40 percent volume fractions.
[b] 16 percent boron fibers volume fraction.
[c] 19 percent boron fibers volume fraction.

3.5.4   E2-Glass

E2-glass parameters include: 9 oz./sq. yd., 38 in. wide, .008 in. thick, 57×54, 8H satin weave. Frequently specified in aerospace applications, this fabric offers excellent strength, formability, and surfacing characteristics. It meets MIL-C-9084C, Type VIII-B specifications. Typical full roll length is 125 yards, 50 in. wide, 8H satin weave.[19]

This fabric is identical to part #543, 7781 E-glass, but is preimpregnated with an epoxy resin system. The material can be stored, shipped, and handled at room temperatures and is cured using a ramp-up schedule requiring, at most, 310°F. Resin content is 27.0 to 33.0 percent.[19]

Tensile strength and density of E- and S-glass are shown in Table 3.11.[20]

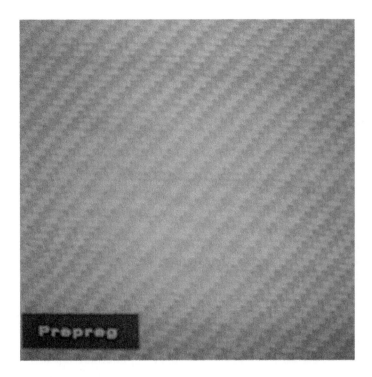

FIGURE 3.4
Prepreg 2×2 twill weave Kevlar®. (From Carbon Fiber Fabrics: www.composites/vision; www.shopmanic.com/carbon. With permission.)

TABLE 3.11

Tensile Strength and Density of E and S Glass

Fiber Type	Tensile Strength, MPa	Density, g/cm³
E-Glass	3,450	2.57
S-Glass	4,710	2.48

3.5.5 Gel Coat

Gel Coat is a high performance, premium quality, gel coat for both airless and air atomized spray applications. No. 180 offers ultraviolet (UV) light stability, excellent clarity, crack, and chemical resistance. This deep, crystal-clear and nonyellowing gel coat is compatible with underwater and water contact marine applications as well as sanitary ware, such as showers and tubs. It is recommended to be used as a protective clear coat over chrome veil, other colors, or glitter flake.

Araldite® 2005 has a minimum curing time and gap filling as well as a simple mixing ratio (2:1 by volume). The curing range is between 20 to 100°C. This provides negligible shrinkage, high shear and peel strength, and heat resistant to 80°C. Araldite 2005 is resistant to water and to a wide range of chemicals.[13,14] Carbon/cyanate ester composites were proposed as replacements for the carbon/epoxy composites with improved out-gassing, microcracking, and moisture absorption characteristics.[15]

3.6   Core Materials: Honeycomb Sandwich Structures and Adhesives

3.6.1   Introduction

The fibers and the matrix are combined in making wind turbine blades. The principal design has selected shells and core materials, which carry proportional loads that are created by the bending and torsion moments. Leading and trailing edges unsupported by inner webs is composed of carbon fibers, with a stiffness two times greater than glass fibers.

3.6.2   Core Materials

Blade manufacturing core materials are used primarily for large-area, unsupported stability in leading/trailing edge panels and shear webs.

Core materials include: end grain balsa, foam cores (PVC, SAN, Urethane, PET-extruded form), engineered core materials (Webcor TYCOR®, NexCore™). End grain balsa has high shear properties and low cost, but is higher in weight. Pet-extruded forms are less cost effective and are lighter.

TYCOR material (fiber-reinforced core) applied to shear webs in the middle of the blade, has very low weight and a high performance core at a highly competitive price. Kitting and assembly in the mold has proved to be faster and more accurate than balsa and foam material.[21]

Fiber reinforced composites consist of closed-cell polyisocyanurate foam (compression strength: 30 kg/m^3), glass fiber reinforcement (E-glass roving mat), and binders.

TYCOR fibers with reinforced core preforms are made with dry fiber and foam. Final mechanical properties of TYCOR are achieved by infusion molding.

A comparison of leading blade core materials is listed in Table 3.12.

Properties of TYCOR W Core include:[21]

- Statistical transverse properties of 25.4 mm TYCOR W molded with Hexion 135i epoxy resin, six-specimen samples
- Comparisons with supplier-published data for:
  - 60 kg/m3 PVC foam (Alcan Airexâ C70.55) average values
  - 155 kg/m^3 balsa (Diab ProBalsaâ PB, "Average" and "Minimum" values)

TABLE 3.12

A Comparison of Leading Blade Core Materials

Core Material	Advantages	Disadvantages
Low density (60-70 kg/m³. Rigid form PVC, SAN	Light weight Affordable at low densities	Low stiffness, more expensive than balsa
Medium density (155 kg/ m3). End grain balsa	Low cost High stiffness	Heavier than foam, subject to processing irregularities, availability problems

Source: Webcore Inc. [21] With permission.

Other options use impregnated prepreg as a honeycomb material.

The axes of honeycomb cells are always quasi-horizontal, and the non-angled rows of honeycomb cells are always horizontally (not vertically) aligned. Thus, each cell has two vertical walls with "floors" and "ceilings" composed of two angled walls. The cells slope slightly upwards, between 9 and 14° toward the open ends.[22]

Two researchers of the geometry of honeycombs offer possible explanations for the reason that a honeycomb is composed of hexagons rather than any other shape. One, given by Jan Brożek, is that the hexagon tiles the plane with minimal surface area. Thus, a hexagonal structure uses the least material to create a lattice of cells within a given volume. Another, given by D'Arcy Wentworth Thompson, is that the shape simply results from the process of individual bees putting cells together—somewhat analogous to the boundary shapes created in a field of soap bubbles. In support of this, he notes that queen cells, which are constructed singly, are irregular and lumpy with no apparent attempt at efficiency.[22]

The closed ends of the honeycomb cells are also an example of geometric efficiency, albeit three dimensional and little noticed. The ends are trihedral (i.e., composed of three planes) sections of rhombic dodecahedra, with the dihedral angles of all adjacent surfaces measuring 120°, the angle that minimizes surface area for a given volume. (The angle formed by the edges at the pyramidal apex is approximately 109° 28' 16" (= 180° - arccos (1/3)). Three-dimension geometry of a honeycomb cell is shown in Figure 3.5.

The shape of the cells is such that two opposing honeycomb layers nest into each other, with each facet of the closed ends being shared by opposing cells (Figure 3.6).

DuPont's aramid fiber honeycomb core PN2 aerospace honeycomb uses include aircraft galleys, flooring, partitions, aircraft leading and trailing edges, missile wings, radomes, antennas, military shelters, fuel tanks, helicopter rotor blades, and navy bulkhead joiner panels.[23]

PN2 honeycomb is available in sheets, blocks, or cut-to-size pieces in both regular hexagonal and over-expanded (OV) cell configurations.

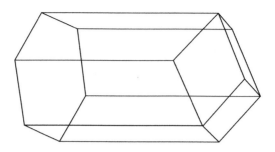

FIGURE 3.5
The three-dimensional geometry of a honeycomb cell. (From Plascore Inc., honeycomb core, online at www.plascore.com/product. With permission.)

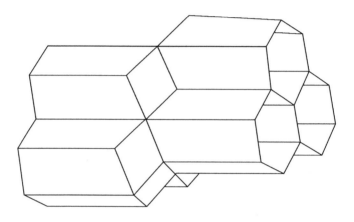

FIGURE 3.6
Opposing layers of honeycomb cells fit together. (From Plascore Inc., honeycomb core, online at www.plascore.com/product. With permission.)

Availability

Cell Sizes: 1/8 - 1/4 in.

Densities: 1.8 pcf - 9.0 pcf

Sheet "Ribbon" (L): 48 in. typical

Sheet "Transverse" (W): 96 in. typical

Tolerances

Length: +3 in., -0 in. (36 in. for OV), width: +6 in. , -0 in.

Thickness: ± .006 in. (under 2 in. thick)

Density: ±10%

Cell Size: ±10%

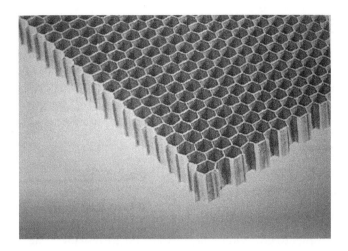

FIGURE 3.7
PN2 aerospace-grade Aramid fiber honeycomb. (From Plascore Inc., honeycomb core, online at www.plascore.com/product. With permission.)

Note: Special dimensions, sizes, tolerances, and specifications can be provided upon request.

PN2 aerospace-grade aramid fiber honeycomb is shown in Figure 3.7.

New prepreg materials have been developed by researchers from NASA, along with Lockheed Martin Engineering.[24] Adhesives studied were AF-555M, XMTA 241/PM15, FM309-1M, and FM-300K.[25-27] The prepreg materials were IM7/MTM45-1 and T40-800B/5320 (Table 3.13).

Adhesives are necessary between structural carbon fibers and honeycombs in prepreg face sheets. 3M adhesives are available as liquids, pastes, tapes, films, and shaped solids. Each has characteristics to be considered for application effectiveness and efficiency.

Pressure-sensitive adhesives (PSAs) can produce lighter structures that are less susceptible to fatigue than those assembled using mechanical attachments. However, the parameters that must be considered to attain optimum performance are very different. To predict how bonds will

TABLE 3.13

New Adhesives

Adhesive	Adhesive Weight, kg/m²	Face Sheet Prepreg
AF-555M	0.39	IM7/MTM45-1
XMTA-241/PM15	0.29	IM7/MTM45-1
FM309-1M	0.39	T40-800B/5320
FM-300K	0.39	T40-800B/5320

withstand the stresses and strains of real applications, it's necessary to look at PSA attributes.[28]

A prestacked, 16-ply prepreg face sheet was laid on a steel tool plate covered with Kapton® film and then sealed. Limitation of vacuum pressure and low viscosity requirements include also fast curing and strength.[29]

3.7   Material Promises a Better Blade Resistance to Wear and Tear

3.7.1   Owens Corning's Ultrablade Fabric Solutions

Owens Corning (Toledo, Ohio) has introduced Ultrablade™ fabric solutions, which will help enable the market transition to longer, lighter, and stiffer wind turbine rotor blades. The new solutions, which were made available in January 2011, can remove nearly a metric ton of reinforcement and resin from 2.0 megawatt (MW) wind turbines compared to same-size blade sets made with traditional E-glass.

Compared to standard fabrics, Ultrablade fabrics in epoxy resin can:

- Reduce spar weight by up to 18 percent while keeping length constant.
- Increase blade length by up to 6 percent.
- Improve blade stiffness by up to 20 percent.
- Decrease blade thickness by up to 6 percent to increase aerodynamic efficiency and generate higher torque for driving turbines.
- Reduce total blade weight by up to 5 percent to ease the load on the turbine and tower, and enable turbines to operate effectively at lower wind speeds.

The fabric is finished with an epoxy resin to establish the strength and rigidity required in a wind turbine blade.

"Ultrablade fabric solutions give designers much more freedom in developing longer blades for today's large turbines," said Dr. Chris Skinner, director of global technical marketing for OCV Technical Fabrics, a division of Owens Corning.

According to Skinner, "Designers can use a combination of several improved properties in different areas of a blade. They can choose to increase blade length for any given weight while keeping the thrust constant and assuring sufficient tower clearance."

Lighter blades also mean turbines will be able to spin and generate electricity with greater efficiency at lower wind speeds.

"At lower wind speeds, weight-saving Ultrablade fabric solutions can help increase a blade's aerodynamic lift, torque, and energy output. The end result will be higher annual energy production from optimized blade designs using high-performance fabrics," according to Skinner.

The company is using distinctive pink stitching to market the product to buyers familiar with the "Pink Panther" Owens Corning fiberglass insulation.

Although turbine blades have been getting longer, they must be sleek, light, and resilient to wear and tear as they spin through the wind. Owens Corning claims that its Ultrablade fabric treated with epoxy resin can help improve blade aerodynamics and strength without sacrificing length.

3.7.2 Film Layer Protects Wind Turbine Blades against Electromagnetic Fields

Special lightning current can generate electromagnetic fields around a wind turbine to which sensitive electronic equipment, e.g., microcontrollers, may be exposed. Thus, the electromagnetic field may cause significant damage to the electronic equipment due to electromagnetic induction of the current in the equipment, which can result in a malfunction of the wind turbine.

The invention[31] relates to a wind turbine comprising a rotating part including a rotor with at least one blade and a wind turbine hub with at least one enclosure structure or similar wall structure, and a stationary part including a nacelle with at least one enclosure structure or similar wall structure. At least one of the parts comprises a conductive film layer of the enclosure structure with connection to a ground potential where the film layer forms a shield enclosing the part or parts and protects against electromagnetic fields. The invention also relates to a method to manufacture the enclosure structure.[30,31]

3.7.3 Painting of Wind Turbines

With the standard painting of wind turbines white, this could be attracting vast numbers of insects, which, in turn, attract birds and bats. A new study has suggested that painting wind turbines purple would stop the blades from attracting and killing thousands of bats and birds each year.[32]

It has long been known that wind turbines attract vast numbers of insects, which in turn bring larger predators within range of the spinning blades. Now, scientists have discovered that the color of the turbines themselves could make a huge difference to the number of insects that are attracted to it. The study, carried out by PhD students from Loughborough University in the United Kingdom, analyzed how the color of a wind turbine influences the number of insects that are attracted to it. They laid colored cards in a random sequence next to a 13 m-high, three-blade wind turbine situated in a meadow near Leicestershire. They found that while most turbines are painted pure white or light grey to blend in with the surroundings, these colors are extremely attractive to insects.

Instead, they suggested that wind turbines painted purple would dramatically reduce the number of insects and, therefore, bats and birds that were brought within range of the spinning blades.

Researcher Chloe Long told the BBC, "Our major conclusion from this work is that turbine paint color could be having a significant impact on the attraction of insect species to the structure, both during the day and at night."

They found turbines painted pure white and light grey drew more insects than any other color apart from yellow.

"We found it extremely interesting that the common turbine paint colors were so attractive to insects," said Miss Long. "If the solution was as simple as painting turbine structures in a different color, this could provide a cost-effective mitigation strategy."

A study last year found that more than 90 percent of birds killed within range of wind turbines actually die because of a sudden change in air pressure before or after each blade, which bursts their blood vessels.

The heat generated by turbines may also play a role in attracting insects and the spinning blades may interfere with bats' echolocation, the researchers said.

The findings are to be published in the *European Journal of Wildlife Research*.

3.8   Field Study of Wind Turbine Blade Erosion

3.8.1   Introduction

Wind turbine blades regularly operate in hostile environments that include severe ice storms, wind with changeable directions, and rainwater. Turbine blades are expected to endure these severe environments without rapid erosion to the leading edge of their rotor blade. To avoid this rapid deterioration and potential irreparable damage, the leading edge is typically protected with erosion-resistant materials.

3.8.2   Field Study and Maintenance

3M (Maplewood, Minnesota) and wind turbine maintenance company, Rope Partner (Santa Cruz, California), announced a new in-field study on how the erosion of the leading edge of wind turbine blades affects power output.[33]

The year-long study will be the first published work to put concrete figures on this kind of damage, which is widespread for wind turbine blades, 3M reported.

Like the edge of a steak knife, the blades of a wind turbine can erode. As they rotate—at speeds of up to 180 mph at the blade tips—outdoor conditions, such as rain, hail, and airborne sand can wreak havoc on the edge, damaging them and, in turn, causing turbulence that negatively affects aero-dynamics and reduces output.

"We've seen firsthand the damage caused by leading edge erosion when conducting our inspections for wind turbine maintenance and repairs," said Chris Bley, director of business development for Rope Partner. "We've seen sites where significant erosion occurs in as little as two years after installation."

The study is important because wind turbine maintenance is costly and, thus, of concern to investors. An unscheduled crane to access failed parts can run $70,000; anything to cut down on emergency maintenance calls is effectively money in the bank.

3M's interest lies in its expertise with polyurethane tape, which it offers for application in the wind power and aerospace industries. One example is its use on helicopter blades.

3.8.3 Polybutadiene Resins

Careful analysis carried out by the "NAT (Neo Advent Technologies LLC) team," identified polybutadiene resins and polyurethane derived from it as a promising material for generating an erosion-resistant polymer framework. Frequent heating of this material for deicing purposes presents the challenging task of having a polymer with high stability at elevated temperatures. This can be addressed with polybutadiene resins.

Polybutadiene resins are a novel class of the hydroxyl-terminated polybu-tadiene homopolymers with a structure that is depicted in Figure 3.8.

They are low molecular weight reactive liquids that offer broad formulating opportunities. Several key attributes are responsible for the formulating advantages provided by polybutadiene resins.

3.8.3.1 Hydroxyl Functionality

Polybutadiene resin end groups are predominantly primary allylic hydroxyl groups (Figure 3.9).[34] These groups have high reactivity with a variety of isocyanates to yield polyurethane polymers. The hydroxyl functionalities of the two widely used grades, polybutadiene R-45HTLO and polybutadiene R-20LM, are typically 2.4 to 2.6 per polymer chain.

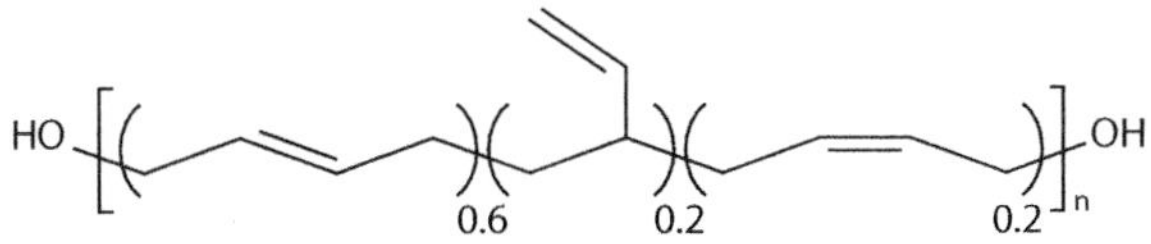

FIGURE 3.8
Chemical structure of the polybutadiene resins ($n = 50$ for polybutadiene R-45HTLO; $n = 25$ for Poly butadiene R-20LM). (From Li, H. 1998. *Rheological behavior of vinyl ester resin*. Online at www.scolar.lib.vt.edu/theses. With permission.)

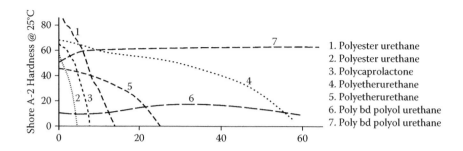

FIGURE 3.9
Comparative hydrolytic stability of the polybutadiene urethane versus other materials. (From Li, H. 1998. *Rheological behavior of vinyl ester resin.* Online at www.scolar.lib.vt.edu/theses. With permission.)

3.8.3.2 Hydrolytic Stability

Polybutadiene resins have a hydrophobic, nonpolar hydrocarbon backbone that imparts hydrolytic stability to products prepared from it. The stability properties surpass those of polyurethanes prepared from other polyols that have ester or ether linkages, which are more hydrophilic and easier to hydrolyze. Polybutadiene resin-based systems can far exceed the 28-day requirement of the U.S. Naval Avionics Center (Indianapolis, Indiana) test. For example, measuring hardness versus time at 100°C and 95 percent relative humidity, it has been shown (see Figure 3.9) that typical polybutadiene-based urethanes are essentially unaffected, whereas urethanes prepared from other polyols actually liquefy (revert) under the test conditions.

The addition of even moderate amounts of polybutadiene resin to polyesterpolyol-based polyurethane markedly improves the hydrolytic stability of the cured polyurethane. Figure 3.10 and Figure 3.11 show test results from two comparable polyurethane systems containing a 24.4 wt. percent polybutadiene resin/75.6 wt. percent polyester polyol mixture and based only on the polyester polyols. Changed tensile strength polybutadiene /polyester/polyurethane composites are shown in Figure 3.10 and changed hardness of polybutadiene/ polyester/polyurethane composites are shown in Figure 3.11.

3.8.3.3 High Hydrophobicity

The core polybutadiene aliphatic chain imparts significant hydrophobic properties to polybutadiene-based urethane as compared to the traditional urethanes. Sand bounces off the fractures upon impact, while raindrops change shape and continue to penetrate the substrate. Therefore, improved hydrophobic properties are essential for the protection from rain erosion

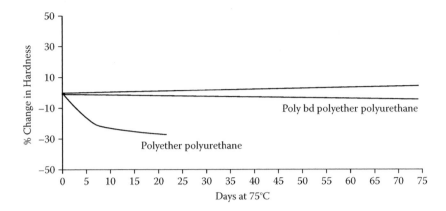

FIGURE 3.10
Changed tensile strength polybutadiene/polyester/polyurethane composites. (From Li, H. 1998. *Rheological behavior of vinyl ester resin.* Online at www.scolar.lib.vt.edu/theses. With permission.)

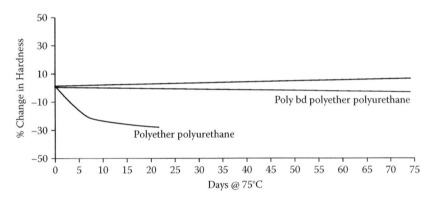

FIGURE 3.11
Changed hardness of polybutadiene polyester/polyurethane composites.

(Figure 3.12). Additionally, more hydrophobic materials will be less prone to icing, whereas deicing procedures may be facilitated.

3.8.3.4 Low Temperature Flexibility

Another attribute of the polyurethane systems based on polybutadiene resins is their outstanding low temperature properties. This characteristic is attributable to the rubbery polybutadiene backbone.

Many polyurethane elastomers derived from polybutadiene resins have brittle points as low as -70°C. This characteristic of the polybutadiene resin-containing formulations also contributes to excellent thermal cycling properties.

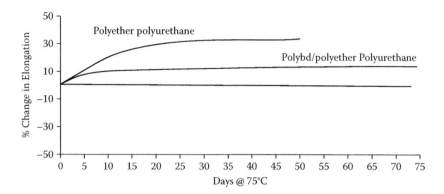

FIGURE 3.12
Change in tensile strength, hardness, and elongation after water immersion.

3.8.3.5 Adhesion Properties

Adhesion strength to both galvanized steel and aluminum is high, characterized by a lap shear within $(758–1172)*10^4$ Pa (American Society for Testing and Materials (ASTM) D102). It changes little with aging, water immersion at room temperature and boiling point, and saltwater immersion.

3.8.4 AIRTHANE PET-91A-Based Elastomers

This approach has been thoroughly validated by the General Electric Company as a part of the program to construct lightweight aircraft engine blades in the late 1990s. Through rigorous trials, a material with superior erosion characteristics was obtained from the prepolymer of the TDI and TMEG, a component commercially supplied by GE's Air Products and Chemicals, Inc. (Bridgeville, Pennsylvania), namely Airthane[a] PET-91A. When cured by Bis-dianiline diamine curative, supplied by Lonza (Basel, Switzerland) under the trademark Lonzacure[a], the resulting elastomer, combined with N-phenylbenzamide (antioxidant) and TINUVIN[a] 765 (hindered amine light stabilizer), has been molded into the blade metal core scaffold. The new composite blade led to significant manufacturing cost reduction, weight savings, and erosion protection properties verified by substantial testing panels. Subsequently, the program has been terminated because of changed corporate priorities and the technology never was commercialized. Some of the properties of the Lonzacure Airthane PET-91A elastomer are given in Table 3.14.[35] Multiple variants of both prepolymer and cure options are commercially available components.

NAT intends to build on the unrealized potential of this group of elastomers based on the commercially available material and enhance their properties for heat conductivity.

Many characteristics are not easily attainable by polyacrylonitrile (PAN)-based carbon fibers. For example, PAN-based fibers have a maximum modulus

TABLE 3.14

Processing Conditions

Lonzacure MCDEA Level, 95% Stoichiometry (%)	17.1
Airthane Temperature (°C)	80
Lonzacure Temperature (°C)	100
Pot Life (Min.) -80 °C	4
Mold Temperature (°C)	130
Demold Time (Min.)	30
Postcure (Hrs./Temp. °C)	48/130

Selected Elastomer Properties

Hardness (A/D)	92/42	Compressive Stress (Pa)	
Modulus (Pa) 100% Elongation	.875*10⁴	5% Deflection	254.5*10⁴
200% Elongation	1,485	10% Deflection	419.3*10⁴
300% Elongation	1,870	15% Deflection	578*10⁴
Tensile Strength (Pa)	3.58*10⁴	20% Deflection	.76*10⁴
Elongation (%)	479	25% Deflection	1*10⁴
Tear Resistance (PLI)	Compressive Set (%)	13.8*10⁴	
Die C	582	Rebound (%)	39.3*10⁴
Split/Trouser	57/86	NBS Abrasion Index	172.4*10⁴

of about 650 Gpa, whereas pitch-based fibers can reach 1,000 GPa. Pitch fibers also are significantly more thermally and electrically conductive. The particular grade of the milled pitch-based carbon fiber proposed for testing is BP/Amoco's ThermalGraph DKD X. This material is graphitized at very high temperatures, which increases thermal and electrical conductivity and modulus of the 28 fibers. Thus, ThermalGraph DKD X, in addition to increased heat conductivity, could improve mechanical stiffness and strength of the resulting composite.

3.8.5 Conclusion

We are developing a technological process for boot manufacturing. We select pultrusion speed process. Polyester fiber drive and impregnate by polybutadiene, polyurethane resin and go through die boot profile. For testing spacemen's (boot prototype) we use four layers.

3.9 Rheological Behavior of Flow Resins

3.9.1 Introduction

Rheology is the study of the flow of materials that behave in an interesting or unusual manner, as in deformation and flow reinforcement behavior.

Rheological behavior of resins has influenced hybrid composites. This behavior involves different phenomena, such as viscosity flow, rubber-like elasticity, viscoelasticity. Such a study is important because viscosity is traditionally regarded as a most important characteristic of resin behavior.

3.9.2   Viscosity

Viscosity is the measurement of a fluid's resistance to flow.[3] For a liquid under shear, the rate of deformation, or shear rate, is proportional to the shearing stress. Newton's law states that the ratio of the stress to the shear rate is constant. That constant is viscosity. The common units of viscosity are poise (P), grams per centimeter second (g/cent.sec), or dyne second per square centimeter (dyn.s/cm^2). The centipoise (cP) one hundred poise is the most common unit.

The SI unit of viscosity is Pascal seconds (Pa.s and mPa.s (1 mPa.s = 1 cP)).[5,36]

Viscosity is an important parameter for the online injection molding process. Epoxy and vinyl ester resins need to have a low viscosity in order to flow into the fibers and to permit good wetting of the carbon fibers. Styrene was used as a diluent to decrease the viscosity of the vinyl ester resins.

Thus, the styrene concentration is an important factor to control the viscosity of resins. Viscosity is highly sensitive to changes in temperature, and the molecular weight of the oligomers also affects viscosity. The effect of temperature and the effect of sterene on viscosity has been investigated.[7,8,37]

3.9.3   Effect of Styrene Contents

The viscosity of the initial resins must be low in order to provide efficient flow of the resin into the fiber preforms in an online molding process. A Brookfield digital viscometer was used to measure the viscosities of molding resins, such as epoxy or vinyl ester. The zero shear viscosity was determined by direct extrapolation of low-shear data from the plot of viscosity versus shear rate. The viscosities were measured for a series of resins (vinyl ester molecular weight = 690 g/mol) with increasing styrene concentrations at room temperature (Figure 3.13).]

The viscosity of vinyl ester decreases with increases in styrene concentration and can be calculated using the equation (3.4):

$$\log \eta = 7.45 - 0.187(\%St) + 0.00143(\%St)^2 \tag{3.4}$$

Vinyl ester resins with varying styrene concentrations provide a system with a wide range of viscosities, which are suitable for infusion molding processes. These resins, with higher styrene concentrations, offer improved resistance to acids and alkalis, while those with lower styrene concentrations afford better resistance to solvent.[37]

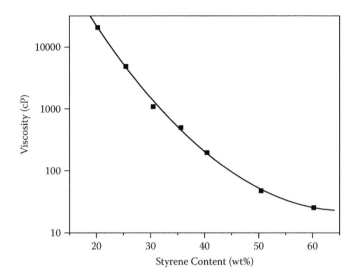

FIGURE 3.13
Viscosity of uncured vinyl ester resins at 25°C (Mn = 690 g/mol).

3.9.4 Effect of Temperature

The viscosity of epoxy or vinyl ester resins decreases as temperature increases. The Andrade equation relates the viscosity to temperature as follows:

$$\eta = A10^{B/T} \tag{3.5}$$

$$\log\eta = \log A + B/T \tag{3.6}$$

where A and B are constants characteristic of the polymer and T is the absolute temperature. The viscosity of a polymer at a given temperature can be estimated by knowing the viscosity at two other temperatures. This knowledge also calculated the constant A and B and determination of viscosities at other temperatures. The Andrade equation holds for many polymer solutions and polymer melts well above their glass-transition temperatures.[6]

The concentrated vinyl ester resins (low styrene content) with two different molecular weights (Mn = 690 g/mol and Mn = 1,000 g/mol) were measured at temperatures of 25°C, 35°C, 50°C, 75°C. Plots of $\log\eta$ versus 1/T are shown in Figure 3.14. It was found that logh versus 1/T was a straight line for all tested resins for temperature ranging from 25 to 75°C, which are well above Tg (glass solidify temperature).

The constants A and B are also determined and listed in Table 3.15. As styrene content increases, the viscosity of the vinyl ester resins become less dependent on temperature, as concluded by the decrease in both log A and B.

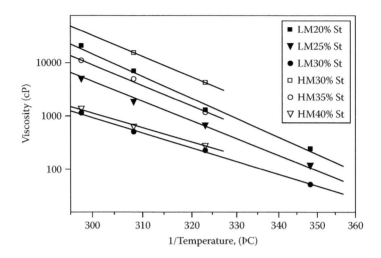

FIGURE 3.14
Plot of log η versus 1/T for different vinyl ester resins.

TABLE 3.15

Constants A and B in the Andrade
Equation for Various Vinyl Ester Resins

Vinyl Ester Resins with Mn = 690 g/mol		
Wt% Styrene	**Log A**	**B**
20	-9.198	3995
25	-8.540	3660
30	-6.89	2983
Vinyl Ester Resins with Mn = 1,000 g/mol		
30	-7.99	3752
35	-8.67	3795
40	-6.13	2758

3.9.5 Effect of Molecular Weight

Unlike dilute solutions, concentrated polymer solutions show a great deal of interaction between the macromolecules. For a concentrated solution, properties above and below Mc, the critical molecular weight for chain entanglement, may be quite different. The dependence of viscosity on molecular weight changes from around unity below Mc to around 3.4 to 3.5 above Mc indicated in equation (3.7) and equation (3.8).[38-40]

$$\eta = KM \quad \text{below Mc} \tag{3.7}$$

$$\eta = KM^{3.4-3.5} \quad \text{above Mc} \tag{3.8}$$

The Mc above which entanglement occurs is usually thought to be on the order of 5,000 to 20,000 g/mol, depending on the chemical structure of the polymer. However, in highly concentrated polymer solutions, dependence of viscosity on molecular weight with exponents larger than 1 may occur at low molecular weights. The viscosity of vinyl ester resin increases significantly when the molecular weight of vinyl ester oligomers increases from 690 g/mol to 1,000 g/mol. For example, at 35 wt.% styrene concentration at room temperature, the viscosity of the cure vinyl ester cure resins (VER) with Mn = 690 g/mol is 521 cP, while that of the VER with Mn = 1,000 g/mol is 1050 cP. The high dependency of viscosity on the molecular weight of the vinyl ester resins of low molecular weight is probably due to hydrogen bonding between the vinyl ester molecules, increasing the effect chain length.

3.9.6   Relations between the Viscosity, Processing, Temperature, and Glass Transition Temperature

The glass transition temperature of the VER with Mn = 690 g/mol was determined by differential scanning calorimetry (DSC). It was found that the vinyl ester resin only have one Tg. The Tg of the vinyl ester resins decreases linearly as styrene concentration increases (Figure 3.15).

For concentrated polymer and oligomer solutions, the Williams–Landel–Ferry (WLF) equation can be used to relate the viscosity with glass transition temperature.[41] In general, the WLF equation holds in the temperature range Tg to (Tg + 100°C):

$$\log \eta / \eta_{Tg} = \frac{-17.4(T - Tg)}{51.6 + (T - Tg)} \tag{3.9}$$

For the vinyl ester/styrene system, a similar relationship exists between the testing temperature, the Tg, and viscosity. In the 20 wt.%, 25 wt.%, and 30 wt.% styrene resins, the plots of $\log \eta$ - (T - Tg) $^{-1}$(Figure 3.16).

Equation (3.7) shows the line fall and viscosity of vinyl ester mixtures are associated with the difference between Tg and the measurement temperature. The curve fits given equation (3.10):

$$\log \eta = \frac{253.2}{T - Tg} - 0.2427 \tag{3.10}$$

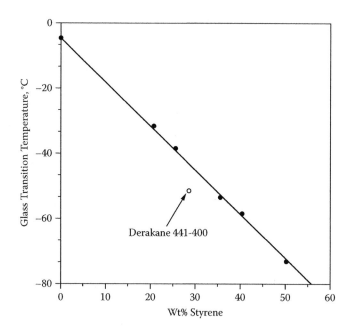

FIGURE 3.15
Glass transition temperature of VER with Mn = 690 g/mol.

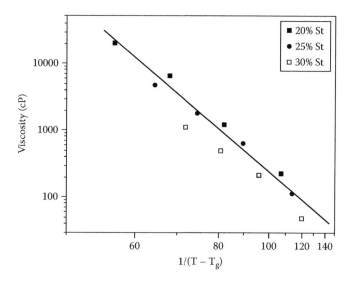

FIGURE 3.16
Plot Log η versus T - Tg for 20 wt.%, 25 wt.%, 35 wt.% styrene of vinyl ester resin with Mn = 690 g/mol.

Because low viscosity in an infusion molding is necessary, the styrene content variation must be carefully tailored to viscosity. The viscosity decreases significantly when temperature increases. On the other hand, good resin stability can only be maintained under 50°C due to the existence of a free radical initiator.

Experimental results in this system suggest that when T - Tg is above 70°C, the low viscosity necessary (under 2,000 cP) for infusion molding processing can be achieved.

References

1. Brondsted, P., H. Lilholt, and A. Lystrup. 2005. *Composite material for wind power turbine blades.* Roskilde, Denmark: Annual Reviews, A Nonprofit Scientific Publisher. email: powlbrondsted@risoe.dk, hans.lilholt@risoe.dk, aage.Lustrup@risoe.dk

2. Lowenstein, K. L. 1993. *The manufacturing technology of continuous glass fibers,* 3rd ed. Amsterdam: Elsevier, pp. 237–291.

3. Hiemenz, P. C. 1984. *Polymer chemistry.* New York: Markcel Dekker.

4. Bikales, N. M., G. C. Overberger, and G. Menges (Eds.) 1988. *Encyclopedia of polymer science and technology,* Vol. 14, New York: John Wiley & Sons.

5. Gaur, B. and J. S. P. Rai. 1993. Rheological and thermal behaviour of vynyl ester resin. *European Polymer Journal* 29 (8): 1149.

6. Scott Blair, G. W. 1969. *Elementary rheology.* New York: Academic Press.

7. Varma, I. K., B. S. Rao, M. S. Choudhary, and D. S. Varma, 1985. Die angewandte makromol. chemistry. *Unsaturated Polyester Technology* 130: 191.

8. Van Wazer, J. R., J. W. Lyons, K. Y. Kim, and R. E. Colwell. 1963. *Viscosity and flow measurement, A laboratory handbook of rheology.* New York: Wiley-Interscience.

9. Li, H. 1998. Rheological behavior of vinyl ester resins Online at: www.scholar.lib.vt.edu/theses

10. Watson, J. C. 2010. Influence of fiberglass sizing design on the performance of composites for wind blades. *SAMPE Journal* 46 (6).

11. Mason, K. F. 2004. Changes in the wind. *Composites Technology,* (April): 26–30.

12. Golfman, Y. 2010, *Hybrid anisotropic materials for structural aviation parts.* Boca Raton, FL: Taylor & Francis Group.

13. Phoenixx TPC, Inc. 2005. Technical data, Taunton, MA.

14. Nuplex Industries. 2010. Technical data. Wadsworth, OH. Online at: www.nuplexconstruction.co.nz/epoxies

15. McConnel, V. P. 1992. *Tough promises from cyanate esters.* Greer, SC: Advanced Composites, May/June.

16. Industry breakthroughs in carbon fiber and aerospace prepregs line. 2010. *Hexcel News, JEC.*

17. Amber Composites. 2010. Datasheet. Online at: www.ambercomposites.com/downloads/datasheet

18. Hahn, G. L. and G. G. Bond. 2011. Non-autoclave (prepreg) manufacturing technology for primary aerospace structures. *SAMPE Journal* 47 (1).
19. Cyntec Engineering Materials. 2010. Cycom®5320-1 datasheet, Online at: http//www.cytec.com/engineered-materials/product/Datasheets/CYCOM%205320-1.pdf
20. Carbon fiber fabrics. Online at: www.fiberglasssupply.com
21. Webcore Inc. Online at: www.webcoreonline.com
22. Honeycomb. Wikipedia, the Free Encyclopedia. Online at: www.en.wikpedia.org/wik/Honeycomb
23. Honeycomb core. Plascore Inc., Beverly Hills, CA. Online at: www.plascore.com
24. Tan-Hung, H., J. M. Baughman, T. J. Zimmerman, J. K. Sutter, J. M. Gardner. 2011. Evaluation of sandwich structure bonding in out of autoclave (OOA) processing. *SAMPE Journal* 47 (1).
25. 3M. *Scotch-Weld™ structural adhesive film AF-555.* Technical Data Sheet. 2007. St. Paul, MN: 3M Aerospace and Aircraft Maintenance Division, June 10.
26. Cytec Engineered Materials. *FM-300 high shear strength modified epoxy film adhesive, and FM 309-1, high performance adhesive film.* Technical Data Sheets. Cytec Engineered Materials, Anaheim, CA, August 16, 2011.
27. Cytec Engineered Materials. *X5320 toughened epoxy for structural applications out of autoclave manufacturing,* Preliminary information sheet, Rev. 1.0–09.03.08. Cytec Engineered Materials, Anaheim, CA, September 3, 2008.
28. 3M. 2010. *Pressure sensitive adhesives in mechanical applications.* 3M Adhesive & Tape Classification. Online at: www.solutions3M.com. January 19, 2011.
29. Kapton Film Tape. 2010. Online at: www.grainger.com/Grainger/Kapton-Film-Tape-4CLE5
30. 3M™ Thermally Conductive Silicone Interface Pads. http://solutions.3m.com/wps/portal/3M/en_US/AdhesivesForElectronics/Home/Products/ThermalSolutions/ThermallyConductiveInterfacePads/
31. Patent 7837443. Wind turbine comprising enclosure structure formed as a Faraday cage. Per Sveigaard Mikkelsen. www.patentgenius.com/patent/7837443. November 23, 2010
32. Firth, N. 2010. Why painting wind turbines purple could protect birds and bats. Online at: www.dailymail.co.uk
33. 3M. Online at: http/www.3M.com
34. Sartomer Company, Inc. 2006. *Sartomer Application Bulletin.* Online at: www.sartomer.com
35. Combat Shield. 2006. Online at: http://www.tapecase.com
36. Bikales, N. M., G. C. Overberger, and C. G. Menges. 1988. *Encyclopedia of polymer science and technology,* Vol. 14 New York: John Wiley & Sons, p. 455.
37. Bruins, P. E. (Ed.) 1976. *Unsaturated Polyester Technology.* New York: Gordon and Breach, p. 315.
38. Vinogradov, G. V., and A. Ya. Msalkin. 1980. *Rheology of Polymers.* New York: Springer-Verlag.
39. Fox, T. G., S. Gratch, and S. Loshaek. 1956. In *Rheology,* ed. F. R. Eirich, Vol. 1. New York: Academic Press.
40. Berry, G., and T. G. Fox. 1968. *Advanced Polymer Science* 5: 261.
41. Williams, M. L., R. F. Landel, and J. D. Ferry. 1955. *Journal American Chemical Society* 77: 3701.

4

Manufacturing Technologies for Turbine Power Blades

4.1 Introduction

Many small companies during the 1970s built wind turbines for energy production. New manufacturing technology was gradually developed during the next two decades (1980s and 1990s). Concern about the working environment and stricter legislation pushed the technologies away from wet and open process toward prepreg technology and closed mold infusion techniques. In this chapter, we describe the (wet) hand lay-up process, filament winding, prepreg technology, and resin infusion technology.

4.2 Wet Hand Lay-Up Process

In the early days, smaller glass fiber, reinforced polyester turbine blades were manufactured using the traditional wet hand lay-up technique in open molds, which had been used for decades in building boats. The reinforcements were all glass fibers and mainly chopped strand mats (CSMs) with random fiber orientation. In some cases, woven fabric was added to increase stiffness and strength. The upper and lower shells were adhesively bonded together to form the airfoil shade blade. As the blade became longer (approaching 8 m), webs were inserted to support the airfoil and to take up both bending and shear loads.

A demand for higher stiffness and strength introduced a more dedicated fiber orientation with more fibers in the longitudinal direction of the blade. This was achieved either by using unidirectional woven fabrics (unbalanced fabrics with more fibers in one of the directions, normally the warp direction) or by laying down many parallel roving fibers in the length direction in between the CSMs. In some designs, these rovings serve another purpose in that they were used to form the root end attachment as well. At the root end, the rovings were wound around steel bushes (small tubes) and continued

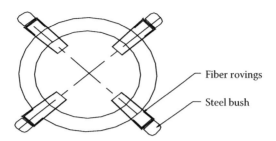

Principal of a hutter flange, illustrating
how the fiber rovings are wound around
steel bushes to form the flange

FIGURE 4.1
Photo of a Hutter flange illustrating how the fiber rovings are wound around steel bushes to form the flange. (From P. Brondsted, H. Lilholt, and A. Lystrup. 2005. *Composite material for wind power turbine blades.* Roskilde, Denmark: Annual Reviews, A nonprofit scientific publisher. With permission.)

back into the blade. The steel bushes formed the holes in what became the flange of the blade to be mounted to the hub. This principal is illustrated in Figure 4.1 and is known as a Hutter flange after its inventor, who used this technique for a wind turbine blade as early as the 1950s.[1] In the mid 1970s, at Tvind in Denmark, a group of pioneers build a three-bladed wind turbine with blade length of 27 m. The blades were manufactured from glass fiber and epoxy in a one-sided open mold. The lower side of the airfoil was laid up in a mold using a combination of CSMs and rovings.

A double Hutter flange with both an external and an internal flange with bolt holes was constructed at the same time.

After curing of the lower airfoil laminate, many transverse foam spars, with an airfoil profile, were placed along the entire length of the blade at equal distances and laminated onto the lower airfoil. Longitudinal webs were integrated as well. The skin laminates of the webs also formed a flange toward the upper airfoil, which was built as the last part of the blade. The Twind turbine is still in operation, but the blades were replaced in 1993.[2]

4.3   Filament Winding

Large wind turbine construction started in the mid 1970s, but then involved companies began looking for more rational manufacturing techniques that were less labor intensive compared with the wet hand lay-up process. Filament winding was a process that was investigated by the several countries.

Filament winding is a sensible way of placing a huge amount of roving around a rotating mandrel in a controlled manner. The shape of wind turbine blade is similar to an aviation profile shape and the majority of the fibers have to be placed along the blade length. Therefore, the filament winding technique had to be developed further for this specific application. In the United States, a technique for winding the entire blade was developed by Kaman Aerospace Corporation and Structural Composite Industries, for blades up to 45 m in length using glass fiber.[3-5] A set of three mandrels was used to gradually build up the airfoil with integrated webs. First, the leading edge part and the forward shear web were wound around the first mandrel, then a second mandrel was attached to the already wound leading edge structure, and more fibers were wound onto the two combined mandrels adding more materials to the first structure and forming a second shear web. Finally a third mandrel was attached, and more winding created the afterbody with the trailing edge. The winding was done with a combination of glass fiber roving and a glass fiber tape; most of the fibers winding in the transverse direction of the tape.

Figure 4.1 illustrates how the fiber rovings are wound around steel bushes to form the flange. This special bias tape was necessary to achieve sufficient bending stiffness and strength of the wind turbine blade. It is difficult and inefficient to place rovings along the axis of mandrels with a filament winding machine. A tape winding technique for winding the spar, including the leading edge and the shear web, was also developed in Denmark by Risoe National Laboratory and the Volund Company in the late 1970s.[6] The blades of the two Nibe wind turbines were 20 m long and were a hybrid construction consisting of a steel spar for the inner 8 m and a glass fiber/polyester spar for the outer 12 m of the blade (Figure 4.2).

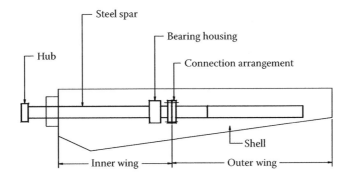

Principal of the nibe blades

FIGURE 4.2
Principal of the Nibe blades: The inner 8 m has a steel spar. (From P. Brondsted, H. Lilholt, and A. Lystrup. 2005. *Composite material for wind power turbine blades*. Roskilde, Denmark: Annual Reviews, A nonprofit scientific publisher. With permission.)

The principal of the Nibe blades is that the inner 8 m has a steel spar and the outer 12 m is an all composite blade.[6]

The airfoil was hand lay-up glass/polyester with a balsa core sandwich in the afterbody, and the airfoil was adhesively bonded to the spars. The thickness of the spar was easily varied along the length from 24 mm at the root end to 6 mm at the tip, by changing the length of travel of the tape placement apparatus along the length of the spar. Because the spar was tapered toward the tip end and the warp glass fibers in the tape could not stretch, it was necessary to insert and withdraw a wedge underneath the tape, at the shear web position of the spar, for each revolution of the mandrel in order to control and maintain straight fibers with the desired orientation in the upper and lower part of the spar. All the slack at the one side of the tape was taken up by the wedge, and the surplus of tape and mismatch of fiber orientation was concentrated at the web where it actually improved the shear properties of the cured laminate. After the final layers of glass fiber tape was wound, a peel ply was applied, and the spar was parked, lined up, and cured.

4.4   Prepreg Technology

The prepreg (preimpregnated) technology has been adapted from the aerospace and aircraft industry. It is based on the used of a semiraw product where the fiber fabrics are preimpregnated by resin and the curing has a portion character.

Golfman[7] demonstrated a way to manufacture sophisticated blades by preliminary curing of spars at 50 to 70 percent. At room temperature, the resin resembles a tacky solid, and the tacky prepregs can be stacked on top of each other to build the designed laminate.

By increasing the temperature, the resin becomes liquid/viscous, and the laminate can be consolidated under pressure and cured into the final component. Prepreg is available in many varieties and combination of fibers, style of fabrics, and resin systems. Different processes of curing temperatures range from 70 to 225°C. For large wind turbine blades, a process temperature of around 80°C is most common.

The required pressure to consolidate the stacked layers of prepregs is achieved by vacuum. The entire lay-up of prepreg is covered by a polymer film, which is sealed to the mold along the edge. A vacuum is pulled underneath the polymer film, and the atmospheric pressure outside the film presses the prepreg package. The shelf life of the prepreg at room temperature normally ranges from a few days to a few weeks depending on the resin system. Therefore, the prepreg is typically stored at −18°C, which results in a shelf life of from 6 to 12 months.

The prepreg technology offers some advantages, such as it is easy to control and to obtain constant material properties, and a higher fiber content gives higher specific stiffness and strength to the material that leads to lighter blades. Also a clean process is obtained, which leads to a better working environment where fewer requirements are needed for the workshop ventilation systems. This results in cost savings on smaller ventilation systems and the energy needed in heating.

Vestas Wind Systems, a Varde, Denmark, wind turbine manufacturer, uses the prepreg technology for its blade production. Vestas has used glass/epoxy prepreg technology for many years, and have now introduced carbon fibers in its 45-m-long blades.[8,9]

4.5   Resin Infusion Technology

Resin infusion technology consists of placing dry fibers in a mold, encapsulating and sealing off the fiber package, injecting the liquid resin into the fiber package, and curing the package. Since the 1990s, the process equipment has been intensively developed, and resin infusion technology is now a widely used industrial process. The final product has not had a dry surface, but is wetted by resins.

Most developments have concentrated on this technique and have tackled it from many sides:

- Fiber sizing with improved wetability to facilitate complete wetting without manipulating the roving fiber fabrics with special architecture to control flow pattern of the resin.
- Resin with low viscosity to improve wetability and lower the process time for large components; resin that does not release volatiles under vacuum.
- Accessories that help control a resin flow pattern over large areas and ensure complete wetting of thick laminates; resin distribution mesh, special sandwich core, etc.
- Equipment for continuous mixing of resin without introducing air
- Design of molds (placement of inlet and outlet for resin, sealing around the edge) to control resin flow to prevent entrapped area with dry fibers.
- Sensors for monitoring the flow front.
- Computer models for the predicting and optimization of the flow pattern for a given component.

LM Glasfiber in Demark is one of the turbine blade manufacturers that uses resin infusion technology, and it has proved to be successful in the company's most recently developed 61.5-m-long blade.[10]

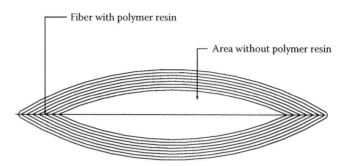

Fiber with polymer resin

Area without polymer resin

Rod section of the large rotorblade

FIGURE 4.3
Setup for resin infusion test of a root end section of a large rotoblade at L. M. Glasfiber, Lunderskov, Denmark.

Generally, it is normal to use a vacuum infusion technique where all the dry fibers are covered with a polymer film sealed to the edge of a mold and held in place against the mold by a vacuum pulled underneath the polymer film (Figure 4.3).

If the resin line is opened, we could pull resin into the dry package with the vacuum. When all the fibers are wetted, the resin line is closed, and the vacuum sucks out surplus resin into a resin trap. The atmospheric pressure outside the polymer film consolidates the laminate toward the mold surface, and the resin is allowed to cure. The two airfoils and the webs are manufactured separately, and all parts are adhesively bonded to complete the blade in a subsequent process.

Siemens Wind Power in Demark has developed a special fiber placement and resin infusion technology for the company's 40-m-long blade, where the fibers for the entire blade, including airfoil, spars, and webs are placed inside a mold cavity and the resin is infused to complete the blade in one pass.[11]

The advantages of resin infusion technologies, compared with those of hand lay-up, are the same as mentioned for the prepreg technology, except for easier process control. The challenge is to control the resin flow to avoid spots with dry fibers in the finished product.

4.6　Out-of-Autoclave Composite Prepreg Process

4.6.1　Introduction

Lightweight composite materials have been widely used in the production of wind turbine blade structures. The airframes of aircraft, such as the F22, F35, and F18, could be designed from reinforced composites. The F35 and others are more costly in dollars if we used autoclave technology. In this book, we concentrate on development of out-of-autoclave new technology, which is less costly and more effective for wind turbine blades. We can infuse resin in a dry package or preliminary impregnate textile fiber and use prepregs to create wind turbine blade structures.

4.6.2 Curing Laminates without Autoclave

The vacuum infusion process has been found to be very versatile for making large composite structures. During the resin impregnation process, the vacuum causes the nucleation and formation of air bubbles from gasses dissolved in the resin.[12] A degassing process, conducted in batches with a small quantity of a bubble nucleation agent (Scotch-Brite™) and an air-sparing method could reduce void formation in the vacuum infusion process. However, despite reducing the level of dissolved gas, some microbubbles remain suspended due to the viscosity of the resin and contribute to void formation. While these results were quite encouraging, there is a need to design a system to ensure a continuous supply of degassed resin as a better way forward. In a continuous degassing system, it is necessary to establish the relationship between the degassing process quality and void content reduction from the perspective of strength and performance, besides signifying the most viable method of degassing.

Preliminary cure impregnates laminate layers at 50 to 70 percent for strength package increase reached by compression molding fiberglass large propeller blades.[7]

Ultrasonic tape lamination (UTL) is a technique developed by Foster Miller, Inc. (Waltham, Massachusetts) to use insonification to achieve heating and consolidation of fiber-reinforced composite materials. In order to induce consolidation without damaging a fiber-reinforced composite, a horn angle of less than 90° is required. Changing the horn angle changes the stress state in the material during insonification and, thus, changes the relative amount of energy dissipated by viscoelastic heating of the matrix compared to fiber disruption.

Foster Miller has demonstrated UTL both for consolidation of thermoplastic matrix composite materials and debulking of B-staged thermoset prepreg. The insonification curing process can induce a very rapid localized heating, the transition glass temperature (Tg) can be achieved in seconds. The viscoelastic properties of the prepreg material at loading rates in the 20 to 40 kHz range are very difficult to measure directly.[13]

Launch vehicles for the U.S. Space-Based Infrared System will include Titan launch vehicles and the Air Force Space Operations Vehicle (SOV). Composite structures on the SOV (the military version of NASA's Reusable Launch Vehicle (RLV)) will be too large to cure inside existing autoclaves. Electron beam (EB) processing is one of the most promising approaches for out-of-autoclave composite curing and bonding.[14] Recent technology demonstration programs have shown potential cost savings and the ability to make large parts using EB curing at a low temperature. However, additional development of EB-cured materials is required to meet RLV and SOV mechanical and thermal design specifications. Science Research Laboratory (Somerville, Massachusetts) and the University of Dayton Research Institute (Dayton, Ohio) have formulated new EB-cured resins and composites and tested these materials at ambient temperature and in liquid nitrogen. The results show that the EB-cured materials developed and the RB-curing process are promising for space structures and for cryogenic applications. EB processing can reduce costs by 10 to 40 percent for production of wind turbine blades. Additional mechanical testing is required to establish a complete set of mechanical properties in liquid nitrogen and in liquid hydrogen. The ultimate goal is to meet or exceed the properties of Cytec Fiberite 977-2, the baseline used by Lockheed Martin in the NASA X-33 Half-Scale Reusable Launch Vehicle program.[15]

The new GKN Aerospace (Redditch, Worcs, United Kingdom) process is called resin film infusion (RFI) and is now being used in conjunction with self heated tooling on the A380 wing trailing edge manufacture. RFI has been shown to offer a 10 percent reduction in manufacturing costs, reduced tooling requirements, lower capital investment costs, lower risk in manufacture, potential weight savings as well as a consistently high quality result.[16] Kubota Research Associates, Inc. (Hockessin, Delaware) has developed a new Green process for manufacturing fiber-reinforced thermoplastic prepregs (PTIR™) out-of–autoclave.[17] The infrared (IR) emitter is dispersed in a thermoplastic matrix. The matrix resin is impregnated onto a reinforcement fabric, such as carbon fiber and polyarilate LCP fiber. PTIR prepreg is made using carbon fiber, and organic fiber in woven fabric and unidirectional forms.

Last year, Boeing Company developed some fast-curing prepreg systems that use a pressure of 101 kPa (14.7 psi) from the vacuum applied to the rubber bag for composite consolidation.[18]

Workable material has a low viscosity when curing at elevated temperatures. Volatile and moisture becomes more critical compared to the autoclave process. Voids created by the entrapment of volatiles are detrimental to the mechanical properties of composite materials. Researchers from NASA and Lockheed Martin Engineering together with BG Smith Associate and Old Dominion University Norfolk developed prepreg packaging IM7/MTM 45-1 and T40-800B/5320[19] using out-of-autoclave (OOA) processing. The technique created a vacuum pressure of 101 kPa (14.7 psi).

4.6.3 Select Technological Parameters and Cure Conditions

Laminate layer consolidation depends on the cure specific pressure, temperature, and time of curing.[20]

Previous research established that low specific pressure of 20 kg/cm^2 did not reduce physical and mechanical properties. However, future reduction of specific pressure increased water penetration between layers and reduced physical and mechanical properties.

The Quickstep proprietary manufacturing process is a range of technologies that can be used in out-of-autoclave fabrication of composite components from advanced composite materials. They were developed and patented by the Australian's Quickstep Technologies Pty. Ltd., with the assistance of the Commonwealth Scientific and Industrial Research Organization, an Australian public sector R&D organization. The balanced-pressure, heated-mold process promises reduced cure cycle times and product weight as well as increased strength and improved appearance. Products produced via the Quickstep process have superior properties to products fabricated by conventional atmospheric cure techniques and have properties generally equal to or better than high-pressure autoclave techniques. The process cycle times are typically 30 to 60 minutes for most resin systems, a significant time savings over the 3 to 8 hours required in autoclave curing processes, while achieving aerospace-grade void contents of less than 2 percent.[21] Significant scrap rate reduction also can be expected, which is otherwise incurred from interrupted cure cycles, as well as reduced capital, tooling, and lower operating costs with energy usage 70 to 90 percent less than for the equivalent autoclave process.

The technology allows large composite parts to be fabricated to aerospace standards together with the added flexibility to co-cure or meld components to produce more structurally integrated complex parts.[12]

The Quickstep process uses a lightweight clamshell-like mold equipped with one or more flexible silicone bladders on each mold side that rapidly applies heat to the enclosed uncured laminate stack.[20] The Ingersoll machine used automated fiber placement and wound a fuselage for Boeing.[22]

The flexible bladders, which permit the laminate to be compressed without subjecting the mold to distortion or stress, have the ability to be rapidly flooded with heat transfer fluid (HTF) so that the mold surface can be heated or cooled far more quickly than is possible with liquid piped inside solid tooling. Three separate tanks are used to contain glycol HTF maintained at three temperatures from cold (room temperature) to hot (up to 400°F). Vibrating HTF applied at constant low pressure (1 psi to 4 psi) against the tool, together with vacuum in the tooling itself, remove air as the part is compacted and cured. Three molding cells (QS1, QS5, QS20) are available that can cure surfaces of 1 to 20 m². The process can accommodate thermoset or thermoplastic prepregs as well as wet resin/dry fiber composite systems. Though best suited to parts having moderate curvature, such as airfoil-shaped components, more complex bladders can be used to mold deeper draw parts. The largest cell, which can be moved by forklift, fits into a 40-ft container and can be installed in one day.

Building on its previous collaboration, Quickstep and the Victorian Centre for Advanced Materials Manufacturing (Belmont, Australia) have initiated a major new R&D program in 2011. The program focuses on composites for aerospace applications, examining particularly how the Quickstep curing technique produces composites with apparent improvements in thermal, adhesion performance, and other properties thought to result from a better cure at the micro and molecular levels. Elsewhere, the National Composites Center in collaboration with the University of Dayton Research Institute, Ashland Performance Materials, Owens Corning, Ohio State University, and WebCore Technologies Inc. will study the Quickstep process using materials enhanced with nanoparticles with the expectation of improving the performance of components produced for the aerospace and automotive industries.

4.7 Developing Technology for Robust Automation Winding Process

4.7.1 Introduction

A robust winding process includes filament fiber or tape preliminary impregnated by resins. Ingersoll's automated fiber placement machine[22] fabricates complex, contoured shapes over large curvilinear surfaces using epoxy-impregnated composite fiber material. The flexibility of Ingersoll's fiber placement process provides the accuracy and performance necessary for advanced aerospace application including fuselage panels, engine cowlings, payload fairings, payload adapters, inlet shrouds, and landing gear pod fairings. Three-dimensional parts are fabricated on form tools supported between centers. Irregular cross sections with an infinitive variety of fiber angles are created by adding or dropping individual fibers to vary band width and achieve localized fiber angles.

4.7.1.1 Fiber Placement Process

Automated Dynamics (Schenectady, New York) manufactures advanced composite structures and high performance composite processing equipment. Through the development of innovative processes—automated fiber placement and automated tape laying—the company delivers strong, lightweight structures that solve complex problems for clients in the aerospace, automotive, defense, oil field, and other commercial/industrial markets.

Filament-wound carbon fiber/fiberglass power turbine blades are provided a very high strength-to-weight ratio and excellent high resistance to corrosion and erosion (Figure 4.4).[22]

Variations in reinforcement materials and resins can be used during the winding process, materials, such as E-glass, C-glass, Kevlar[a], and carbon fiber as reinforcement. Resins, such as epoxy and vinyl esters with low viscosity, demonstrate a high life capability and minimum vocabulary.

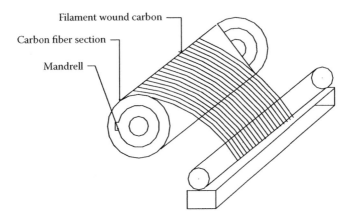

Sketch of Ingersoll's automated fiber placement machine

FIGURE 4.4
Ingersoll's automated fiber placement machine. (From Ingersoll Machine Tools. Online at www.ingersoll.com. With permission.)

The fiber placement process utilizes narrow-width tapes, typically 0.25 to 1 in. (6–25 mm), as its composite material medium, and lays down heats via hot gas, laser, or other heating methods. This consolidates the composite material onto the tool, *in situ*, without the need for further consolidation processes (see Figure 4.4).

An automation composite structure process provided by Automated Dynamics is shown in Figure 4.5.

4.7.1.2 Continuous Molding Prepreg Process

The continuous molding prepreg process is advantageous for blade turbine winds.

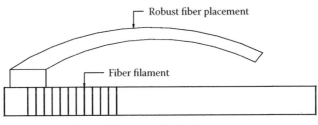

Automation composite structure process
provided by automated dynamics

FIGURE 4.5
Automation composite structure process provided by Automated Dynamics. (From Ingersoll Machine Tools. Online at www.ingersoll.com. With permission.)

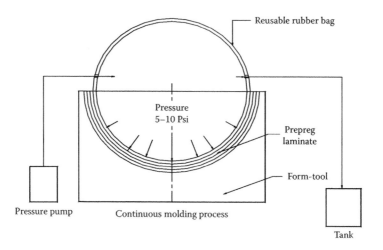

FIGURE 4.6
Continuous molding process. (From www.en_wikipedia.org/wiki/rotational_molding.)

The prepregs are applicable to out-of-autoclave and the curing process is the same used in SCRIMP (Seeman's composite resin infusion molding process): vacuum by pump and reusable bags. Fiberglass and carbon is preliminarily impregnated with epoxy or vinyl ester or liquid polymers. Then, silicon bags are rapidly fitted to the infusion lines; in this case, they are only fitted to vacuum lines and bags, which improves the repeatability of the process. For curing process and cooling, use standard air lines. Setup time and process robustness greatly improve when using molding or pultrusion spar caps.

The continuous molding process is shown in Figure 4.6. Pressure pumps insert air into the rubber/silicon bag and pressurized to 5 to 10 psi. This creates a rubber reusable bag. This pressure is compressed into a laminate impregnated package. After 10 min., pressure is released and the process is continuous in the second area.[24]

After installing spar caps in the bottom laminates, we use hydraulic power hinges for spin top laminates on bottom laminates (Figure 4.7). Hinges eliminate flip fixtures bottom and top, which improves accuracy and greatly reduces assembly time.[23,24]

4.8 Infusion Molding Process

4.8.1 Introduction

Seeman Composite, Inc. (SCI) and its predecessor TPI Company have been involved in the reinforced plastics industry since the 1990s. Bill Seeman

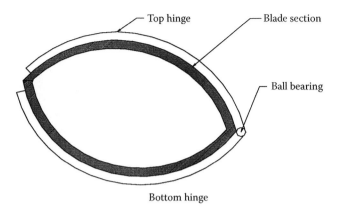

Hydraulic power hinges for blade assembly

FIGURE 4.7
Hydraulic power hinges for blade assembly.

recognized the labor intensiveness, health risks, and environmental impact of traditional hand-laid open molding laminating methods in the 1970s and began experimenting with vacuum infusion. The product of this research and development was SCRIMP, the Seeman's composites resin infusion molding process. This technology illuminates VOC (volatile organic compound) emissions and creates a clean, worker-friendly environment. It also produces aerospace-grade mechanical properties with a practical size limit, low cost tooling, and consistent results. SCRIM was awarded its first US Patent since 1990. Its patent has been used throughout the world to fabricate a wide variety of high-quality composite products for military and commercial clients, including pleasure boats, surface and subsurface naval applications, ground vehicles, and aircraft structures. The success of SCRIM was a result of having a background in commercial composites and an operation for aerospace quality. The first customer was the Naval Surface Warfare Center, Carderock Division, for whom it built a test module for the advanced technology composite deckhouse program. When the project proved an unqualified success, the navy began to accept that there was a way to build high-quality composite parts at a price that was affordable for large-scale structures on navy ships. The infusion process of the thick laminate stack has been used to absorb 15 drums of resin that was dry less than seven hours, and then was cured quietly under vacuum overnight. The resin automation infusion process has been identified as a cost-effective fabrication technique for producing tolerant textile composites. Dry textile preforms have been resin impregnated, consolidated, and cured in a single step, eliminating costly prepreg-type manufacture and ply-by-ply layup. The large number of material properties of solid processing parameters must be specified and controlled during resin infiltration and cure of textile composites. The development of a

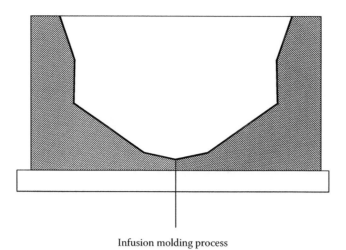

Infusion molding process

FIGURE 4.8
Seeman's composite resin infusion molding (SCRIM) process.

comprehensive three-dimensional process simulation model for fabrication of varying stiffened structures are described in references 22,23.

The carbon/glass epoxy film process gives an equal resin distribution and has the automation control for the solid-dip prepreg impregnation.[23] Prepreging and stitching a fabric to ensure fiber alignment adds a cost making the economics less palatable. The high strength-to-weight ratio of carbon allows for smaller-diameter blade root connections similar to a glass-only design. In SCRIM, a strong vacuum is used to saturate, or infuse, a dry layup with resin at a single-sided (male or female) mold. SCRIM built wind blades with ratios of 70:30 fiber-to-resin with 1 percent void content and a low VOC emission, along with mechanical properties that rival those associated with prepregs or wet lay-up and autoclave cure (Figure 4.8).

Several new forms of vacuum-molding may save FRP (styrene regulation) boat builders from the regulators. (The new fabricated SCRIM process is partly owned by production sailboat builder TPI Composites, Inc.) A strong vacuum draws catalyzed resin into a dry laminate stack to saturate the fiber reinforcement. Because the part also cures under the bag, this system virtually eliminates styrene emission.[25,26]

Prepreg impregnation is an alternative and more automated process, but it is not used for wind turbine blade fabrication today. This process was investigated in works by Ahn and Seferis[27,28] and Lee, Seferis, and Bonner.[29] Consistency evaluation of a qualified fiber prepreg system was studied by Buehler, Seferis, and Zeng.[30] Hayes and Seferis[31] researched the self-adhesive honeycomb prepreg systems for secondary structural applications. Braiding and RTM succeed in aircraft primary structures is shown as alternative technology.[32] However, Virginia Tech College of Engineering demonstrated resin

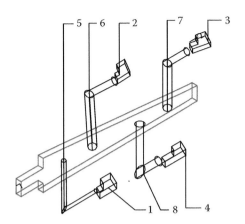

1–4 Vacuum pumps
5-Piping section 1
6-Piping section 2
7-Piping section 3
8-Piping section 4

Sequential sections transportation resin transfer molding

FIGURE 4.9
Sequential sections transport resins by a transfer molding process. Sequential sections under the vacuum bag moves the resin from the vacuum pumps, pos. 1–4; piping pos. 5–8 distributes a resin into area.

film infusion (RFI) process simulation of complex wing structures.[33] The impregnation process for prepregs and braided composites was published by Golfman.[34] Engineering and manufacturing robust solutions changes in the wind was described by Mason in Industry Composite Technology.[35] Figure 4.9 shows a mold-mounted resin detection system and a superior control computer technology used to optimize the gate opening.

TPI Inc. uses the infusion molding process by transferring epoxy resin directly by vacuum pumps and impregnating the fiber (Figure 4.10).

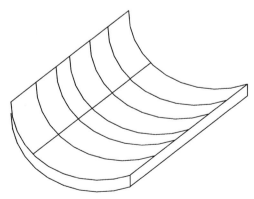

Anatomy of wind turbine blade (near max chord)

FIGURE 4.10
Shows the transfer epoxy resin in half of mold turbine blade.

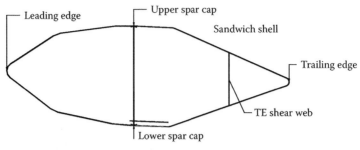

Wind turbine blade components

FIGURE 4.11
Anatomy of wind turbine blades (near max chord).

The anatomy of wind turbine blades (near max chord) is shown in Figure 4.11.

The carbon spar consists of an upper spar cap, inner shear web, and lower spar cap, which are manufactured by a continuous molding process. When inner shear web is main spar endorse and gave strength, It is main spar endorse strength when acting bending and twisting forces acted in longitudinal directions. Robust SCRIM process, reduce cost and time (Figure 4.12).

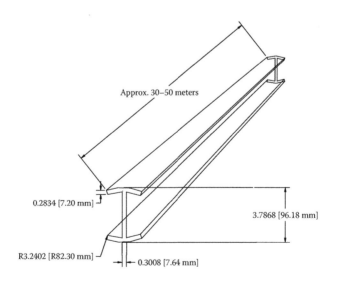

Carbon spar for large-scale turbine

FIGURE 4.12
Carbon spar for large-scale turbine blade.

Cetex(r) Thermo-Lite(r) TC1200 is a semicrystalline polyetherketone thermoplastic composite UD tape that can be used for a longitudinal spar (70 percent carbon, 30 percent thermoplastic). Mechanical properties for Thermo-Lite TC1200 are listed in Table 4.1 and Table 4.2.

For many reasons, hybrids, combining glass fiber with carbon, have more advantages for strength in wind turbine blades.[30]

TABLE 4.1

Mechanical Properties–CETEX Thermo-Lite TS1200 PEEK AS-4.

Property	Condition	Test Method	Result
Tensile Strength(0°)	RTD	ASTM D3039	330 ksi (2280 MPa)
Tensile Modulus(0°)	RTD	ASTM D3039	20.4 Msi (141 GPa)
Poisson's Ratio	RTD	ASTM D3039	0.329
Tensile Strength (90°)	RTD	ASTM D3039	9.68 ksi (66.5 MPa)
Tensile Modulus (90°)	RTD	ASTM D3039	1.3 Msi (9.0 GPa)
Compressive Strength(0°)	RTD	ASTM D3039	159 ksi (1100 Mpa)
Compressive Modulus(0°)	RTD	ASTM D3039	17.7Msi (122 GPa)
Compressive Strength(0°)	ETD[a]	ASTM D6641	172 ksi (1193 MPa)
Compressive Strength(0°)	ETW[b]	ASTM D6641	147 ksi (1013 MPa)
In Plane Shear Strength V-Notch[c]	RTD	ASTM D5379	12.1 ksi (83.4 MPa)
In Plane Shear Modulus V-Notch[d]	RTD	ASTM D5379	1.03 Msi (7.10 GPa)
Compression Modulus(0°)	ETW	ASTM D3410	17.1 Msi (118 GPa)
Interlaminar Shear Strength (SBS)	RTD	ASTM D2344	12.9 ksi (88.9 MPa)
Interlaminar Shear Strength (SBS)	ETW	ASTM D2344	10.8 ksi (74.2 MPa)
Open Hole Tensile Strength	RTD	ASTM D5766	56.1 ksi (387 Mpa)
Open Hole Compressive Strength	RTD	ASTM D6484	46.1 ksi (318 MPa)
Double Bearing Shear Strength	RTD	ASTM D5961	155 ksi (1070 MPa)

Source: TenCate.[36]

[a] ETD is tested at 250°F/121°C.

[b] ETW is tested at 180°F/82°C after 14 days soaking in 160°F/71°C water.

[c] Measured at 5 percent offset.

[d] In-Plane shear strength V-notch.

TABLE 4.2

Mechanical Properties–CETEX Thermo-Lite TS1200 PEEK 1M-7

Property	Condition	Test Method	Result
Tensile Strength (0°)	RTD	ASTM D3039	400 ksi (2750 MPa)
Tensile Modulus (0°)	RTD	ASTM D3039	25.0 Msi (172 GPa)
Tensile Strength (90°)	RTD	ASTM D3039	12.5 ksi (86.2 MPa)
Tensile Modulus (90°)	RTD	ASTM D3039	1.5 Msi (10.3 GPa)
Compressive Strength (0°)	RTD	ASTM D6641	160 ksi (1103 Mpa)
Interlaminar Shear Strength (SBS)	RTD	ASTM D2344	16.0 ksi (110 MPa)

Source: TenCate.[36]

4.8.2   Conclusion

The automatic, 3 axial braiding process with the automation VARTM process and Smart Molding Control were proposed for use for wind blades manufacturing.

4.9   Rotational Molding

4.9.1   Introduction

Rotational molding, also known as rotomolding, rotocasting, or spin casting, is a molding process for creating many kinds of mostly hollow items, typically of plastic, that can be used for molding web spars for wind turbine blades.

A three-motor-powered (tripower) rotational molding or spin casting machine was developed recently. This is a robust manufacturing process for web spars for wind turbine blades.

A heated, hollow mold is filled with a charge or shot weight of material. It is then slowly rotated (usually around two perpendicular axes) causing the softened material to disperse and stick to the walls of the mold. In order to maintain even thickness throughout the part and to avoid sagging or deformation during the cooling phase, the mold continues to rotate at all times during the heating phase. The process was applied to plastics in the 1940s, but, in the early years, it was rarely used because it was a slow process restricted to a small number of plastics. Over the past two decades, improvements in process control and developments with plastic and ceramics powders have resulted in a significant increase in usage. (For a description of basic principles of rotational molding, see Wikipedia.[37])

4.9.2   History

In 1855, R. Peters of Britain documented the first use of biaxial rotation and heat. This rotational molding process was used to create metal artillery shells and other hollow vessels. The main purpose of using rotational molding was to create consistency in wall thickness and density. In 1905, in the United States, F. A. Voelke used this method for the hollowing of wax objects. This led to G.S. Baker's and G.W. Perks' process of making hollow chocolate eggs in 1910. Rotational molding developed further and R. J. Powell used this process for molding plaster of Paris in the 1920s. These early methods using different materials directed the advancements in the way rotational molding is used today with plastics.[38]

Plastics were introduced to the rotational molding process in the early 1950s. One of the first applications was to manufacture doll heads. The machinery was made of an E Blue box-oven machine, inspired by a General

Motors rear axle, powered by an external electric motor and heated by floor-mounted gas burners. The mold was made out of electroformed nickel–copper, and the plastic was a liquid PVC plastisol. The cooling method consisted of placing the mold into cold water. This process of rotational molding led to the creation of other plastic toys. As demand and popularity of this process increased, it was used to create other products, such as road cones, marine buoys, and car armrests.

This popularity led to the development of larger machinery. A new system of heating was also created, going from the original direct gas jets to the current indirect high velocity air system. In Europe during the 1960s, the Engel process was developed. This allowed large hollow containers to be created of low-density polyethylene. The cooling method consisted of turning off the burners and allowing the plastic to harden while still rocking in the mold.[39]

In 1976, the Association of Rotational Molders (ARM) was started in Chicago as a worldwide trade association. The main objective of this association was to increase awareness of the rotational molding technology and process.[39]

In the 1980s, new plastics, such as polycarbonate, polyester, and nylon, were introduced to rotational molding. This has led to new uses for this process, such as the creation of fuel tanks and industrial moldings. The research that has been done since the late 1980s at Queen's University Belfast in Northern Ireland has led to the development of more precise monitoring and control of the cooling processes based on their development of the "rotolog system."

4.9.3   Equipment and Tooling

Rotational molding machines are made in a wide range of sizes. They normally consist of molds, an oven, a cooling chamber, and mold spindles. The spindles are mounted on a rotating axis, which provides a uniform coating of the plastic inside each mold.[40-43]

Tooling is either fabricated from welded sheet steel or cast.[39] The fabrication method is often driven by part size and complexity; most intricate parts are likely made out of cast tooling. Molds are typically manufactured from stainless steel or aluminum. Aluminum molds are usually much thicker than an equivalent steel mold, as it is a softer metal. This thickness does not affect cycle times significantly since aluminum's thermal conductivity is many times greater than steel. Due to the need to develop a model prior to casting, cast molds tend to have additional costs associated with the manufacturing of the tooling, whereas fabricated steel or aluminum molds, particularly when used for less complex parts, are less expensive. However, some molds contain both aluminum and steel. This allows for variable thicknesses in the walls of the product. While this process is not as precise as injection molding, it does provide the designer with more options. The aluminum addition to the steel provides more heat capacity, causing the melt-flow to stay in a fluid state for a longer period.

4.9.4   Standard Setup and Equipment for Rotational Molding

Normally all rotation molding systems have a number of parts including molds, cooling chamber, and mold spindles. The molds are used to create the parts, and are typically made from aluminum. The quality and finish of the product is directly related to the quality of the mold being used. The oven is used to heat the part while also rotating the part to form the part desired. The cooling chamber is where the part is placed until it cools, and the spindles are mounted to rotate and provide a uniform coat of plastic inside each mold.

4.9.4.1   *Rock and Roll Rotating Molding Machines*

This is a specialized, single-arm machine. It rotates or rolls the mold 360° in one direction and at the same time tips and rocks the mold 45° above or below horizontal in the other direction. Newer machines use forced hot air to heat the mold. These machines are best for large parts that have long length-to-width ratio. Because of the smaller heating chambers, there is a savings in heating costs.[43]

4.9.4.2   *Clamshell Machine*

This is a single-arm, rotational molding machine. The arm is usually supported by other arms on both ends. The clamshell machine heats and cools the mold in the same chamber. It takes up less space than equivalent shuttle and swing arm rotational molders. It is low in cost compared to the size of products made. It is available in smaller scales for schools interested in prototyping and for high-quality models. More than one mold can be attached to the single arm.[41,43]

4.9.4.3   *Vertical or Up and Over Rotational Machine*

The loading and unloading area is at the front of the machine between the heating and cooling areas. These machines vary in size between small to medium compared to other rotational machines. Vertical rotational molding machines are energy efficient due to their compact heating and cooling chambers. These machines have the same (or similar) capabilities as the horizontal carousel, multiarm machines, but take up much less space.[42,43]

4.9.4.4   *Shuttle or Swing Arm Machine*

This is a single-arm, turret machine that moves the mold back and forth between the heating and cooling chambers. This machine moves the mold in a linear direction in and out of heating and cooling chambers. It is low in cost for the size of product produced. It is also available in smaller scale for schools and prototyping

4.9.4.5 Carousel Machine

This is one of the most common machines in the industry. It can have up to six arms and comes in a wide range of sizes. The machine comes in two different models: fixed and independent. A fixed carousel consists of three fixed arms that must move together. One arm will be in the heating chamber while the other is in the cooling chamber and the other in the loading/reloading area. The fixed carousel works well when working with the same mold. The independent carousel machines are available with more arms that can move separately from the others. This allows for different size molds, with different heating and thickness needs.[43]

4.9.5 Production Process

The rotational molding process is a high-temperature, low-pressure plastic-forming process that uses heat and biaxial rotation (i.e., angular rotation on two axes) to produce hollow, one-piece parts. Critics of the process point to its long cycle times—only one or two cycles an hour can typically occur, as opposed to other processes, such as injection molding, where parts can be made in a few seconds. The process does have distinct advantages. Manufacturing large, hollow parts, such as oil tanks, is much easier by rotational molding than any other method. Rotational molds are significantly cheaper than other types of molds. Very little material is wasted using this process, and excess material can often be re-used, making it a very economically and environmentally viable manufacturing process (Figure 4.13).

The rotational molding process consists of four distinct phases:

1. Loading a measured quantity of polymer (usually in powder form) into the mold.
2. Heating the mold in an oven while it rotates until all the polymer has melted and adhered to the mold wall. The hollow part should be rotated through two or more axes, rotating at different speeds, in order to avoid the accumulation of polymer powder. The length of time the mold spends in the oven is critical, too long and the polymer will degrade, reducing impact strength. If the mold spends too little time in the oven, the polymer melt may be incomplete. The polymer grains will not have time to fully melt and coalesce on the mold wall, resulting in large bubbles in the polymer. This has an adverse effect on the mechanical properties of the finished product.

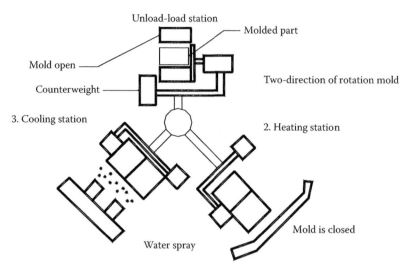

Rotational molding process

FIGURE 4.13
Rotational molding process.

3. Cooling the mold, usually by fan. This stage of the cycle can be quite lengthy. The polymer must be cooled so that it solidifies and can be handled safely by the operator. This typically takes tens of minutes. The part will shrink on cooling, coming away from the mold, and facilitating easy removal of the part. The cooling rate must be kept within a certain range. Very rapid cooling (e.g., water spray) would result in cooling and shrinking at an uncontrolled rate, producing a warped part.

4. Removal of the part.

4.9.6   Recent Improvements

Until recently, the process was largely empirical, relying on both trial and error and the experience of the operator to judge when the part should be removed from the oven, and when it was cool enough to be removed from the mold. Technology has improved in recent years, allowing the air temperature in the mold to be monitored, removing much of the guesswork from the process.

Much of the current research is into reducing the cycle time as well as improving part quality. The most promising area is in mold pressurization. It is well known that applying a small amount of pressure internally to the mold at the correct point in the heating phase accelerates coalescence of the polymer particles during the melting, producing a part with fewer bubbles in less time than at atmospheric pressure. This pressure delays the separation of the part from the mold wall due to shrinkage during the cooling phase, aiding cooling of the part. The main drawback to this is the danger to the operator of explosion of a pressurized part. This has prevented adoption of mold pressurization on a large scale by rotomolding manufacturers.

4.9.7 Mold Release Agents

A good mold release agent (MRA) will allow the material to be removed quickly and effectively. Mold releases can reduce cycle times, defects, and browning of the finished product. There are a number of mold release types available; they can be categorized as follows:

- Sacrificial coatings: The coating of MRA has to be applied each time because most of the MRA comes off on the molded part when it releases from the tool. Silicones are typical MRA compounds in this category.
- Semipermanent coatings: The coating, if applied correctly, will last for a number of releases before requiring reapplication or touching up. This type of coating is most prevalent in today's rotational molding industry. The active chemistry involved in these coatings is typically a polysiloxane.
- Permanent coatings: Most often some form of PTFE (polytetrafluoroethylene) coating, which is applied to the mold. Permanent coatings avoid the need for operator application, but may become damaged by misuse.

4.9.8 Materials

More than 80 percent of all the material used is from the polyethylene family: cross-linked polyethylene (PE), linear low-density polyethylene (LLDPE), and high-density polyethylene (HDPE). Other compounds are glass and carbon fiber, PVC plastisols, nylons, and polypropylene and ceramics. The order of materials most commonly used by industry is:

- Polyethylene
- Polypropylene
- Polyvinyl chloride
- Nylon
- Polycarbonate

These materials are also occasionally used (not necessarily in order of most used):

- Aluminum
- Acrylonitrile butadiene styrene (ABS)
- Acetal
- Acrylic
- Epoxy
- Fluorocarbons
- Ionomer
- Polybutylene
- Polyester
- Polystyrene
- Polyurethane
- Silicone–ceramics mixture

4.9.8.1   Natural Materials

Recently it has become possible to use natural materials in the molding process. Through the use of real sands and stone chip, sandstone composite can be created, which is 80 percent natural nonprocessed material.

4.9.9   Products

Designers can select the best material for their application, including materials that meet U.S. Food and Drug Administration (FDA) requirements. Additives for weather resistance, flame retardation, or static elimination can be incorporated. Inserts, threads, handles, minor undercuts, flat surfaces without draft angles, or fine surface detail can be part of the design.

Designs can also be multiwall, either hollow or foam filled. Products that can be manufactured using rotational molding include storage tanks, bins and refuse containers, airplane parts, doll parts, road cones, footballs, helmets, rowing boats, and kayak hulls. Playground slides and roofs are also generally rotomolded. Nacelle shell can be produced by prepreg fiberglass, using a rotation molding process. We can use injection molding, employing reinforced liquid polymers. There is a method of making a wind turbine blade by rotate molds.[44]

4.9.9.1   Product Design

There are many considerations for a designer when designing a part. Which factors are most important to a client? For instance, a part may need to be cheap and a certain color. However, if another color is cheaper, would the client be willing to change colors? Designers are responsible for considering all the limitations and benefits of using certain plastics. This may result in a new process being decided upon.

4.9.9.2 Designing for Rotational Molding

Another consideration is in the draft angles. These are required to remove the piece from the mold. On the outside walls, a draft angle of 1° may work (assuming no rough surface or holes). On inside walls, such as the inside of a boat hull, a draft angle of 5° may be required. This is due to shrinkage and possible part warping.

Another consideration is of structural support ribs. While solid ribs may be desirable and achievable in injection molding and other processes, a hollow rib is the best solution in rotational molding. A solid rib may be achieved through inserting a finished piece in the mold, but this adds cost.

Rotational molding excels at producing hollow parts. However, care must be taken when this is done. When the depth of the recess is greater than the width, there may be problems with even heating and cooling. Additionally, enough room must be left between the parallel walls to allow for the melt-flow to properly move throughout the mold. Otherwise webbing may occur. A desirable parallel wall scenario would have a gap at least three times the nominal wall thickness, with five times the nominal wall thickness being optimal. Sharp corners for parallel walls also must be considered. With angles of less than 45° bridging, webbing and voids may occur.

4.9.9.3 Material Limitations and Considerations

Another consideration is the melt-flow of materials. Certain materials, such as nylon, will require larger radii than other materials. Additionally, the stiffness of the set material may be a factor. More structural and strengthening measures may be required when a flimsy material is used.

4.9.9.4 Wall Thickness

One benefit of rotational molding is the ability to experiment, particularly with wall thicknesses. Cost is entirely dependent on wall thickness, with thicker walls being costlier and more time consuming to produce. While the wall thickness can be nearly any thickness, designers must remember that the thicker the wall, the more material and time will be required, increasing costs. In some cases, the plastics may significantly degrade due to extended periods at high temperatures. Also, different materials have different thermal conductivity, meaning they require different times in the heating chamber and cooling chamber. Ideally, the part will be tested to use the minimum thickness required for the application. This minimum will then be established as a nominal thickness.

For the designer, while variable thicknesses are possible, a process called *stop rotation* is required. This process is limited in that only one side of the mold may be thicker than the others. After the mold is rotated and all the surfaces are sufficiently coated with the melt-flow, the rotation stops and the melt-flow is allowed to pool at the bottom of the mold cavity.

Wall thickness is important for corner radii as well. Large outside radii are preferable to small radii. Large inside radii are also preferable to small inside radii. This allows for a more even flow of material and a more even wall thickness. However, it is to be noted that an outside corner is generally stronger than a inside corner.

4.9.10 Process: Advantages, Limitations, and Material Requirements

Rotational molding offers design advantages over other molding processes. With proper design, parts assembled from several pieces can be molded as one part, eliminating high fabrication costs. The process also has inherent design strengths, such as consistent wall thickness and strong outside corners that are virtually stress free. For additional strength, reinforcing ribs can be designed into the part. Along with being designed into the part, they can be added to the mold.

The ability to add prefinished pieces to the mold alone is a large advantage. Metal threads, internal pipes and structures, and even different colored plastics can all be added to the mold prior to the addition of plastic pellets. However, care must be taken to ensure that minimal shrinkage while cooling will not damage the part. This shrinking allows for mild undercuts and negates the need for ejection mechanisms (in most pieces).

Another advantage lies in the molds themselves. Since they require less tooling, they can be manufactured and put into production much more quickly than other molding processes. This is especially true for complex parts, which may require large amounts of tooling for other molding processes.

Rotational molding is also the desired process for short runs and rush deliveries. The molds can be swapped quickly or different colors can be used without purging the mold. With other processes, purging may be required to swap colors.

Due to the uniform thicknesses achieved, large stretched sections are nonexistent, which makes large thin panels possible (although warping may occur). Also, there is little flow of plastic (stretching), but rather a placing of the material within the part. These thin walls also limit cost and production time.

Another cost-limiting factor is the amount of material wasted in production. There are no sprues or runners (as in injection molding), no off-cuts (thermoforming), or pinch off scrap (blow molding). What material is wasted, through scrap or failed part testing, can usually be recycled.

4.9.10.1 Limitations

Rotationally molded parts have to follow some restrictions that are different from other plastic processes. As it is a low-pressure process, sometimes designers face hard to reach areas in the mold. Good quality powder may help overcome some situations, but usually the designers have to keep in mind that it is not possible to make some sharp threads used in injection molded goods. Some products based on polyethylene can be put in the mold before filling it with the main material. This can help to avoid holes that otherwise would appear in some areas. This could also be achieved using molds with movable sections.

Another limitation lies in the molds themselves. Unlike other processes where only the product needs to be cooled before being removed, with rotational molding, the entire mold must be cooled. While water cooling processes are possible, there is still a significant down time of the mold. Additionally, this increases both financial and environmental costs. Some plastics will degrade with the long heating cycles or in the process of turning them into a powder to be melted.

4.9.11 Conclusions

1. Due to high temperatures within the mold, the plastic must have a high resistance to permanent change in properties caused by heat (high thermal stability).

2. The melted plastic will come into contact with the oxygen inside the mold; this can potentially lead to oxidation of the melted plastic and deterioration of the material's properties. Therefore, the chosen plastic must have a sufficient amount of antioxidant molecules to prevent such degradation in its liquid state.

3. Because there is no pressure to push the plastic into the mold, the chosen plastic must be able to flow easily through the cavities of the mold.

4. The part's design must also take into account the flow characteristics of the particular plastic chosen. Rotational molding process for creating many kinds of mostly hollow items, typically of plastic, can be used for molding web spars wind turbine blades.

References

1. Brondsted, P., H. Lilholt, and A. Lystrup. 2005. *Composite material for wind power turbine blades*. Roskilde, Denmark: Annual Reviews, A Nonprofit Scientific Publisher. email: powlbrondsted@risoe.dk, hans.lilholt@risoe.dk, aage.Lustrup@risoe.dk

2. Fakta om Vindenergi. 2003. Faktablade M5. Danish Wind Turbine Owners Association. Online at: http/www.dkwind.dk/fakta/fakta..pdf/M5.pdf

3. Kaman Aerospace Corp. 1976. *Design study of wind turbines 50 kW to 300 kW for electric utility application.* Kaman Rep. No. R-1382, ERDA/NASA-19404-76/2. Bloomfield, CT: Kaman Aerospace Corporation, pp. 4-5, 4-51, 4-79.

4. White, M. L., and W. D. Weigel. 1979. A low cost composite blade for a 300-foot diameter wind turbine. Paper presented at the 34th Annual Technical Conference of the Reinforced Plastics/Composites Institute. Sec. 15-C: 1-6, Soc. Plastic Ind., Washington, D.C.

5. Weingard, O. 1979. Fabrication of large composite spars and blades. Paper presented at the 34th Annual Technical Conference of the Reinforced Plastics/Composites Institute. Sec. 15-B: 1-4, Soc. Plastic Ind., Washington, D.C.

6. Johansen, B. S., H. Lilholt, and A. Lystrup. 1980, Wing blades of glass fiber reinforced polyester for a 630 kW wind turbine—Design, fabrication and material testing. Paper presented at the Third International Conference on Composite Materials, Paris, Aug. 26–29.

7. Golfman, Y. 1970. Manufacturing sophisticated blades by preliminary curing spars at 50–70 percent. USSR Patent 263860.

8. Online at: http://www.vestas.com/producter/pdf/updates..020804/v90..3.UK.pdf

9. Griffin, D. A. 2004. Growing opportunities and challenges in wind turbine blade manufacturing. High-performance composites. Online at: http://www.composites world.com/hpc/issues/204/May/450

10. Glasfiber. Lunderskov, Denmark. Online at: http://www.lm.dk

11. Griffin. D.A. 2004. Cost and performance tradeoffs for carbon fibers in wind turbine blades. *SAMPE Journal* (June/July): 20–28.

12. Afendi, M., W. M. Banks, and D. Kirkwood. 2005. *Bubble free resin for infusion process.* Glasgow, Scotland: University of Strathclyde, SAMPE.

13. Player J., M. Roylance, et al. 2000. UTL Consolidation and Out-of Autoclave Curing of Thick Composite Structures. Paper presented at the 32nd Annual Conference of ISTC, Boston, November 5–9. Online at: http://web.mit.edu/roylance/www/sampe00.pdf

14. Byrne, C. 2000. *Non-autoclave materials for large composite structures.* Somerville, MA: Science Research Lab. Online at: http://www.stormingmedia.us/89/8974/A897483.html

15. Tatum, S. 2001. Lockheed Martin demonstrates low cost method for manufacturing large complex composite structures in advanced fleet ballistic missile project. Press release. Lockheed Martin, Bethesda, MD. Online at: http://www.lockheedmartin.com/wms/findPage do?dsp

16. GKN Aerospace develops manufacturing processes for complex composite structures. July 2006. Online at: http://www.azom.com/details.asp?newsID=6054

17. PTIR™ prepreg and composite. 2000. Hockessin, DE: Kubota Research Association, Inc. Online at: www.ir-welding.com

18. Hou, T. H., J. M. Baughman, T. J. Zimmerman, J. K. Sutter, and J. M. Gardner. 2011. Evaluation of sandwich structure bonding in out–of-autoclave (OOA) processing. *SAMPE Journal* 47 (1).

19. Bahareva, B. E. 1968. Fiberglass in shipbuilding industry. *Shipbuilding Journal.*

20. Coenen, V., M. Hatrick, H. Law, D. Brosius, A. Nesbitt, and D. Bond. 2010. A feasibility study of quickstep process of an aerospace composite material. Perth, Australia: Quickstep Technologies Pty Ltd. www.quickstep.com.au/files/document/21_Feasibility_Study_Quickstep_SAMPE2005.pdf. Sept. 8.

21. Golfman, Y. 1971. Manufacturing large details from fiberglass without using hydraulic presses. *Shipbuilding Technology* 4.

22. Ingersoll. Automated fiber placement. 2009. Online at: http://www.ingersoll.com/fiber.com, Jan. 3.

23. U.S. Department of Energy. 2010. 2.20 percent wind energy by 2030. July 20. Online at: www.20percentwind.org

24. Nolet, S. C. 2010. Manufacturing of utility-scale wind turbine blades. Warren, RI: TPI Composites, Inc. Online at: www.iawind.org/presentations/nolet.pdf

25. Lazarus, P. 1994. Infusion. *Professional Boatbuilder* 31.

26. Lazarus, P. 1997. Reporting from the resin infusion front. *Professional Boatbuilder* 44 (December/January).

27. Ahn, K. J., and J. C. Seferis 1993. Prepreg process analysis. *Polymer Composites* 14 (4): 346.

28. Ahn, K. J., and J. C. Seferis. 1993. Prepreg process science and engineering. *Polymer Engineering and Science* 33 (18): 1177.

29. Lee, W. J., J. C. Seferis, and D. C. Bonner. 1986. Prepreg processing science. *SAMPE Quarterly* 17 (2): 58.

30. Buehler, F. U., J. C. Seferis, and S. Zeng. 2001. Consistency evaluation of a qualified glass fiber prepreg system. *Journal of Advanced Materials* 34 (2): 41.

31. B.S. Hayes and J.C. Seferis, "Self-adhesive honeycomb prepreg systems for secondary structural applications," *Polymer Composites* 19 (1)54(1998).

32. Stover, D. 1994. Braiding and RTM succeed in aircraft primary structures. *High Performance Composites* (January/February).

33. MacRae, D. 1995. *Resin film infusion (RFI) process simulation of complex wing structures*. Blacksburg, VA: Virginia Tech College of Engineering. Online at: http://www.sv.vt.edu/comp_sim/macrae.html

34. Golfman, Y. 2007. Impregnation process for prepregs and braided composites. *SAMPE Journal* (special edition) 3.

35. Mason, K. F. 2004. Changes in the wind. *Engineering & Manufacturing Solutions to Industry Composite Technology* (April).

36. TenCate. 2010. Technical data. Online at: www.tencate.com

37. Wikipedia, the free encyclopedia. 2010. Rotation molding. Online at: en.wikipedia.org/wiki/rotational_molding

38. Ward, N. M. 1997. A history of rotational moulding. *Platiquarian Reprints*. (Archived from the original in 2009.)

39. Beall, G. 1998. *Rotational molding*. Cincinnati, OH: Hanser Gardner Publications, pp. 18, 62–68, 69–77, 152–155.

40. Todd, R. H., D. K. Allen, and L. Alting. 1994. *Manufacturing processes reference guide*. New York: Industrial Press Inc. Online at: http://books.google.com/books?id=6x1smAf_PAcC

41. Crawford, R, and J. L. Throne. 2002. *Rotational moulding of plastics*. Norwich, NY: William Andrew Inc.

42. Crawford, R, and M. Kearns. 2003. *Practical guide to rotational moulding*. Shropshire, U.K.: Rapra Technology Ltd.

43. Online at: http://machinedesign.com/article/putting-the-right-spin-on-rotational-molding-designs-0518
44. U.S. Patent Application 20100122459. 2010. Method of making wind turbine blade. www.patentstorm.us/applications/20100122459.description.html. May 20.

5

Dynamic Strength

5.1 Stress and Vibration Analysis of Composite Wind Turbine Blades

5.1.1 Introduction

Wind turbine blades work under extreme and changeable wind loads, especially in winter with ice storms and during the summer with hurricanes and rain storms. Displacements of blades depend on the stiffness of fiberglass and deformations that create stresses and possible failure. We can analyze this environmental situation by using the laminate theory.

Composite propeller blades work similar to helicopter rotors and fans, which have been manufactured in a laminate form from fiberglass and carbon fiber, reinforced with epoxy. The blade skin also is made with aramid fiber, which also has been reinforced with epoxy resin. Additional reinforcement is necessary due to the impact of birds on the leading edge, which is made from thermoplastics to protect the laminated layers.

The carbon fibers with reinforced epoxy resin have been oriented in various directions. Changing the fiber direction in the layers can turn the torsion and bending stiffness of composite blades.

5.2 Stress Analysis of Propeller Blades

The natural frequencies of propeller turbine blades can be favorably placed in an area outside the operational rpm range. This following relates the evaluation of the force and free vibration frequencies for the purpose of avoiding the noise air and ground resonance. In the early 1970s, composite propeller blades with reinforced epoxy were utilized throughout the world.[1] Propeller blades and fans were manufactured from fiberglass, which was protected by special polymers.[2] For impact protection, the leading edge was covered by thin copper inputted in the prepreg package during compression molding.

The history of the manufacturing process is very significant because it relates how the configuration and architecture of design was determined. A new chapter in aviation history opened in 2003 with the maiden flight of the world's first civil tilt rotor, the Bell/Augusta Aerospace BA609.[3] The rotors were installed in the vertical position and hovered at an altitude of 50 feet, performed left and right peddle turns, both forward and aft flight maneuvers and four take-off and landings. When the rotors are tilted forward in the horizontal position, the aircraft is able to fly as a turboprop fixed-wing airplane. The transition from helicopter mode to airplane mode takes 2 sec, as does the transition from airplane mode to helicopter mode. For the design and manufacturing of the prototype YUH-60A UTTAS tail rotor, the pultrusion process[4] was selected. Fully cured pultruded spar design layout and manufacturing approach was developed. This is important since the pultruded spar consists of the two blades being manufactured simultaneously. Elastic coupling, which has a significant effect on the dynamic elastic torsion response of the rotor, was reported by E. C. Smith.[5]

We have assumed that the propeller blade, as presented in Figure 5.1, is a curved plate with variable thickness and forces acting on it. Acting forces include: air stream load, rotating forces, and bending and torsion moments. I have assumed that the blade rotates about the Z axis, with a constant angular velocity ω (see Figure 5.1). We have X, Y, and Z Cartesian coordinates and polar coordinates R, θ, and Z.

In Figure 5.1, we see a cut through the equilibrium element of a rotor blade and the normal and shear stress distribution in the top section.

Here, Fz is axial force, Fx, Fy are tangential forces and Mz, My, Mx are rotating moments relative to axes Z, Y, X. The area of cross section is denoted

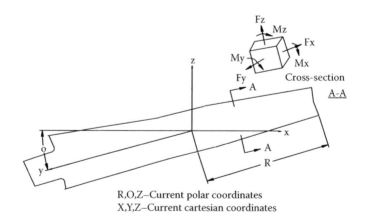

R,O,Z–Current polar coordinates
X,Y,Z–Current cartesian coordinates

Wind turbine rotor blades

FIGURE 5.1
Wind turbine rotor blades.

by A-A and the volume of the region above this plane, by V, while ρ is designated the mass density, ω the angular velocity, r is the radius of rotating blade, and the components of the force acting on the cross section A-A are given by:

$$Fx = \rho\omega^2 \int_V \cos\theta \, r^2 x \, dr \, d\theta \, dx \qquad (5.1)$$

$$Fx = Fy = \rho\omega^2 \int_V \cos\theta \, r^2 x \, dr \, d\theta \, dx \qquad (5.2)$$

While the moments about the x, y, z axis are:

$$M_x = \rho \, \omega^2 \int r^2 \sin\theta \, dr \, d\theta \, dx \qquad (5.3)$$

$$M_y = \rho \, \omega^2 \int x \, r^2 \cos\theta \, dr \, d\theta \, dx \qquad (5.4)$$

$$M_z = \rho \, \omega^2 \int x \, r^2 \sin\theta \, dr \, d\theta \, dx \qquad (5.5)$$

Here:
θ is the angle of the rotating blade.
Equation (1.1) to Equation (1.5) can be solved as:

$$Fz = -\rho\omega^2 x \, r^3/3 \, \cos\theta \qquad (5.6)$$

$$Fx = Fy = \rho\omega^2 x \, r^3/3 \, \sin\theta \qquad (5.7)$$

$$M_x{}^R = -\rho \, \omega^2 x \cos\theta \, r^3/3 \qquad (5.8)$$

$$M_y{}^R = -\rho \, \omega^2 x^2 \cos\theta \, r^3/6 \qquad (5.9)$$

$$M_z{}^R = -\rho \, \omega^2 x^2 \sin\theta \, r^3/6 \qquad (5.10)$$

5.3 Theoretical Investigation

In the linear performance, relationship between the acting forces and the linear deformation in the matrix forming the following was established:

Fz,Fy,Fx	Q_ε	$Q_{\varepsilon X}$	$Q_{\varepsilon R}$	$Q_{\varepsilon\theta}$	ε	
M_z^R	$Q_{X\varepsilon}$	Q_X	Q_{XR}	$Q_{X\theta}$	Υ_Z	(5.11)
M_y^R	$Q_{R\varepsilon}$	Q_{Rx}	Q_R	$Q_{R\theta}$	Υ_R	
M_x^R	$Q_{\theta\varepsilon}$	$Q_{\theta x}$	$Q_{\theta y}$	Q_θ	Υ_θ	

Here: R, θ, Z are the current polar coordinates; $\varepsilon, \Upsilon x, \Upsilon_R, \Upsilon_\theta$ are the current deformation existing from the acting forces Fz, Fy, Fx and bending moments Mz, My, and torsion moment Mx. Q_{ki} is the coefficient of stiffness in the polar coordinates and $Q_{ki} = Q_{ik}$ (k, I = ε, x, R, θ).

This system follows the theory of Kerhgofa–Klebsha and has been transformed on five independent relationships represented in:

$$\varepsilon z = \frac{Fz}{E_1 S_1}; \varepsilon y = \frac{Fy}{E_2 S_2}; \Upsilon_R = \frac{My^R}{E_1 I_R}; \Upsilon_\theta = \frac{MX^R}{G_{xz} T_x}$$

Here:

E_1, E_2 are the modulus of elasticity;
S_1, S_2 are the sections area ;
I_z, I_R are the moments of inertia for axes x, R;
G_{xz} is a shear module;
T_x is a geometrical stiffness for torsion;
T_x is a moment of inertia for axes x.

These parameters can be determined in:

$$I_z = \int y^2 dS; I_R = \int z^2 dS;$$

The geometrical stiffness for torsion can be determined as:

$$T_x = \int_S (R^2 \theta dS$$

If we assume that the stress components in the top section can be found using Equations designated in the polar coordinates, we get:

$$\sigma_r = \frac{1}{r} \times \frac{d\Phi}{dr} + \frac{1}{r^2} \times \frac{d^2\Phi}{d\theta^2} + Z(r,\theta)$$

$$\sigma_\theta = \frac{d^2\Phi}{dr^2} + Y(r,\theta) \tag{5.12}$$

$$\tau_{r\theta} = \frac{1}{r^2} \times \frac{d\Phi}{d\theta} - \frac{1}{r} \times \frac{d^2\Phi}{drd^2} = \frac{d}{dr}\left(\frac{1}{r} \times \frac{d\Phi}{d\theta}\right)$$

Here: Φ is the stress function and a function of variables r and θ.

We also assume that $Z(r,\theta) = Y(r,\theta) = -\int R dr$ where R is a body force $R_r = Fz$ and $R_\theta = Fy$; therefore, Equation (5.12) will result in the following:

$$\sigma_r = \frac{1}{r} \times \frac{d\Phi}{dr} + \frac{1}{r^2} \times \frac{d^2\Phi}{d\theta^2} - \int Fz(r,\theta)$$

$$\sigma_\theta = \frac{d^2\Phi}{dr^2} - \int Fy(r,\theta) \tag{5.13}$$

$$\tau_{r\theta} = \frac{1}{r^2} \times \frac{d\Phi}{d\theta} - \frac{1}{r} \times \frac{d^2\Phi}{drd\theta} = \frac{d}{dr}\left(\frac{1}{r} \times \frac{d\Phi}{d\theta}\right)$$

The stress function Φ seen in Chou and Pagano[6] can be found as Equation (5.14):

$$\Phi = M\,\Psi \tag{5.14}$$

Here, coefficient M has a constant value and Ψ is a geometrical profile function.

In Golfman,[7] the aviation profile Ψ is represented by Equation (5.15).

$$\lambda x^2 - k(y-\alpha)(y-\beta) = 0 \tag{5.15}$$

The coefficient of profile K can be found as:

$$K = \frac{27}{16} \frac{e_k^2 \lambda}{\alpha - \beta} \tag{5.16}$$

where:

e_k is a maximum length of profile;

α is a distance from exit x to exit edge;

β is a distance from exit x to entry edge;

λ is the coefficient of anisotropy.

Therefore,

$$\lambda = \frac{G_{yz}}{G_{xz}} \text{ represents the equation for the skin layers;} \tag{5.17}$$

G_{yz}, G_{xz} = modulus of shear in xz, yz interlaminar directions.

For a composite carbon/fiber epoxy sandwich structure, the modulus of shear can be determined as Equation (5.18).

$$\lambda = \frac{G_{xz}^s + G_{yz}^h}{G_{xz}^s + G_{yz}^h} \tag{5.18}$$

The shear modulus with the index s can be designated for the skin layers; the shear modulus with index h can be designated for the sandwich layers.

The functional relationship between Cartesian and polar coordinates can be shown in Equation (5.19).

$$x = r\cos\theta; \ y = r\sin\theta; \ r^2 = x^2 + y^2;$$

$$\frac{dr}{dx} = \frac{x}{r} = \cos\theta; \ \frac{dr}{dy} = \frac{y}{r} = \sin\theta; \ \frac{d\theta}{dx} = \frac{y}{r^2} = \frac{\sin\theta}{r}; \frac{d\theta}{dy} = \frac{x}{r^2} = \frac{\cos\theta}{r} \quad (5.19)$$

If we replace x and y in Equation (5.19), the contour of the aviation profile can be shown as the following equation:

$$\Psi = \lambda r^2 \cos^2\theta - k(r\sin\theta - \alpha)\ (r\sin\theta - \beta) = 0;$$

The stress function Φ from Equation (5.14) can be shown in polar coordinates as:

$$\Phi = M(\lambda r^2 \cos^2\theta - k(r\sin\theta - \alpha)\ (r\sin\theta - \beta)$$

The complementary equation between Cartesian and polar coordinates, Equation (5.20), would follow:[6]

$$\frac{d^2\Phi}{dr^2} + \frac{1}{r}\frac{d\Phi}{dr} + \frac{1}{r^2}\frac{d^2}{d\theta^2} = \frac{d^2\Phi}{dx^2} + \frac{d^2\Phi}{dy^2} \quad (5.20)$$

The left side of the complementary Equation (5.20) represents the normal radial stresses acting in the r direction, while the outside forces represent a combination of tension forces Fz, Fy, and bending moments M_z^r and M_y^r.

From the integrating stress function $\Phi = M\Psi = M[\lambda x^2 - k(y-\alpha)\ (y-\beta)]$, which is relative to x and y, we get Equation (5.21).

$$\frac{d^2\Phi}{dx^2} = M[2\lambda - k(y - \alpha)(y - \beta)] ; \ \frac{d^2\Phi}{dy^2} = M[\lambda x^2 - 2k + k(\beta + \alpha - \alpha\beta)] \quad (5.21)$$

Therefore, by adding this function, we can find coefficient M (see Equation (5.22)).

$$M[2\lambda - k(y - \alpha)(y - \beta)] + [\lambda x^2 - 2k + k(\beta + \alpha - \alpha\beta)] = Fz + Fy + Mz + My$$

$$M = \frac{Fz + Fy + Mz + My}{[2\lambda - k(y - \alpha)(y - \beta)] + [\lambda x^2 - 2k + k(\beta + \alpha - \alpha\beta)]} \quad (5.22)$$

We already know the stress function Φ in polar coordinates (Equation (5.14)), so we can find all the components necessary to determine the normal and shear stresses, Equation (5.23).

$$\frac{d\Phi}{dr} = 2M\left[\lambda r \cos^2\theta - k(\sin\theta - \alpha)(\sin\theta - \beta)\right];$$

$$\frac{d^2\Phi}{dr^2} = 2M\left[\lambda \cos^2\theta - k(\sin\theta - \alpha)(\sin\theta - \beta)\right];$$

$$\frac{d\Phi}{d\theta} = 2M\left[2\lambda r \cos\theta - k(\sin\theta - \alpha)(\sin\theta - \beta)\right];$$

$$\frac{d^2\Phi}{dr^2} = 2M\left[-2\lambda r \sin\theta + k(\sin\theta - \alpha)(\sin\theta - \beta)\right];$$

(5.23)

Now, for the normal and shear stresses we find (Equation (5.24)):

$$\sigma_r = 2M(\lambda\cos^2\theta - 2\lambda/r\,\sin\theta - \rho\omega^2 r/3\cos\theta)$$

$$\sigma_\theta = 2M[\lambda\cos^2\theta - k(\sin\theta - \alpha)(\sin\theta - \beta)] - \rho\omega^2 r^3/3\sin\theta \qquad (5.24)$$

$$\tau_{r\theta} = 2M[2\lambda\cos\theta - k/r(\cos\theta - \alpha)(\cos\theta - \beta)]$$

In the case of torsion rotor blades, the equation of compatibility can be shown as Equation (5.25).

$$\frac{d\tau_{xz}}{dy} - \frac{d\tau_{xz}}{dx} = (-2Gxz - 2Gyz)\Upsilon \qquad (5.25)$$

Here, the shear stresses are represented as Equation (5.26).

$$\tau_{xz} = \frac{d\Phi}{dy}; \quad \tau_{yz} = \frac{d\Phi}{dx} \qquad (5.26)$$

The stress function is represented in Equation (5.14).

In this case, the necessary components look like:

$$\frac{d\Phi}{dy} = M_1[\lambda x^2 - k(2y - \beta - \alpha + \alpha\beta)] \qquad (5.27)$$

By substituting Equation (5.27), we find coefficient M_1:

$$M_1 = \frac{(-2Gxz - 2Gyz)\Upsilon}{\lambda x(x-2) - k[y^2 - \alpha(1+y) - \beta(1+y) + 2\alpha\beta]} \qquad (5.28)$$

5.4   Vibration Analysis

The longitudinal motion of a stable helicopter will be found to exhibit two modes, which are damped oscillations. The first mode with light damping and a relatively long period is called the *long-period* or *phugoid* mode. The second heavy damped mode is referred to simply as the *short-period* mode. The equation for forced vibration without damping is:

$$m\frac{\partial^2 \varpi}{\partial x^2} + Q_{11}x = (L - D - W\sin\theta)\sin\Omega t \qquad (5.29)$$

Here:

Ω is a forcing frequency acting from turbulence movement;
t is a time of a wave propagation;
m is the mass of the blade, is constant all the time and doesn't depend upon movement and attitude.

We assume a periodic force of magnitude; $F = (L - D - W\sin\theta)\sin\Omega t$.
Here:
L is a lift force;
D is a drag wind force;
W is a weight of a rotor blade.

The force of the lift is actually much stronger than the wind's force against the front side of the blade, which is called *drag*. In the case of free vibration when turbulence movement doesn't exist: $(L-D-W\sin\theta)\sin\Omega t = 0$, which has the following solution as:[8]

$$x = C_1\sin\varpi t + C_2\cos\varpi t \qquad (5.30)$$

and where circular frequency $\varpi = (Q_{11}/m)^{1/2}$.
Here, Q_{11} is the stiffness of the rotor blade and can be determined using previous equations. C_1 and C_2 are arbitrary constants.
We assume $C_1 = A\cos\phi$; $C_2 = A\sin\phi$. Here, A is an amplitude of vibration and ϕ is a phase angle of vibration. We input this into Equation (5.30) and the

displacement x will be:

$$x = A\cos\phi\,\sin\varpi t + A\sin\phi\,\cos\varpi t \qquad (5.31)$$

or $x = A\sin(\varpi t + \phi)$. We replace circular frequency $\omega = 2\pi f$, where f is a motion frequency, so:

$$x = \sin(2\pi f + \phi) \qquad (5.32)$$

and:

$$f = \frac{1}{2\pi}\left(\frac{Q_{11}}{m}\right)^{1/2} \qquad (5.33)$$

By differentiating Equation (5.33), we can determine velocity and acceleration of the longitudinal vibration:

$$V = (2\pi f + \phi)\cos(2\pi f + \phi) \qquad (5.34)$$

$$a = -(2\pi f + \phi)^2 \sin(2\pi f + \phi) \qquad (5.35)$$

The equation for force vibration with damping in the longitudinal direction becomes:

$$m = \frac{\partial^2 \varpi}{\partial x^2} + \delta\frac{\partial \varpi}{x} + Q_{11}x = \left(L - D - W\sin\theta\right)\sin\Omega t \qquad (5.36)$$

Here:
  m is a mass of a rotor blade;
  δ is a critical damping coefficient.

The particular solution that applies to the steady-state vibration of the helicopter should be a harmonic function of time, such as:[9]

$$x_p = A\sin(\Omega t - \phi) \qquad (5.37)$$

where A and ϕ are constant.
  Substituting x_p in Equation (5.37), we get:

$$-m\Omega^2 A\sin(\Omega t - \phi) + \delta\Omega A\cos(\Omega t - \phi) + Q_{11}A\sin(\Omega t - \phi) = (T - D - W\sin\theta)\sin\Omega t \qquad (5.38)$$

Submitting two boundary conditions $\Omega t - \phi = 0$ or $\Omega t - \phi = \pi/2$ results in:

$A(Q_{11} - m\Omega^2) = (L - D - W \sin \theta) \sin\Omega t$ and $\delta\Omega A = (L - D - W \sin \theta) \sin\Omega t$ (5.39)

The phase angle ϕ reflects a different phase between the applied force and the resulting vibration and is determined as:

$$\tan \phi = \frac{\delta\Omega}{Q_{11} - m\Omega^2}$$ (5.40.)

The sine and cosine function can be eliminated:

$$A\delta\Omega + A(Q_{11} - m\Omega^2) = (L - D - W \sin \theta)$$ (5.41)

The forcing frequency Ω can be determined if $\tan\phi = 1$ as:

$$\Omega = (Q_{11}A - (L - D - W \sin \theta) / mA)^{1/2}$$ (5.42)

The forcing frequency Ω has never been fit with a natural frequency f, which avoids parametric resonance. Submitting two boundary conditions:

$$\Omega t - \phi = 0; \text{ or } \Omega t - \phi = \pi/2 \text{ results in:}$$
$$\delta\Omega A = (L - D - W \sin \theta) \sin\Omega t$$ (5.43)
$$-m\Omega^2 t^2 A + Q_{11}A = (L - D - W \sin \theta) \sin\Omega t$$

Thus, the magnitude of amplitude changes from A_{1m} to A_{2m}:

$$A_{1m} = \frac{(L - D - W\sin\theta)\sin\Omega t}{\delta\Omega}; \quad A_{2m} = \frac{(L - D - W\sin\theta)\sin\Omega t}{-m\Omega^2 t^2 + Q_{11}}$$ (5.44)

The electrical circuit for electronic countermeasures to compensate vibration and the critical damping coefficient δ for the carbon–epoxy composite of the blades can be determined as the relationship between the potential energy, W, and the energy lost during one deformation cycle, dW.

$$\delta = \frac{dW}{W} = m\omega\lambda = m2\pi f^{1-v}\lambda$$ (5.45)

Here:
   m is the mass of a helicopter;
   ω is the natural circular frequency, $\omega = 2\pi f^{1-v}$;
   λ is the coefficient of internal friction;
   f is the frequency of the cycle of variation of the deformation;
   v is the exponent and is dependent on the frequency f.

According to Bok, v = 0, while according to Fokht , v = 1.[10] Fokht's hypothesis concerning the proportionally of the nonelastic stress to the frequency is not confirmed by experiment, while the Bok's hypothesis is in better agreement with experimental results, at least in a rather wide range of frequencies.

Golfman[8] shows the critical damping coefficient, δ, for fiberglass for different angles relative to a warp/fill directions. The critical damping coefficient also has been determined in the process of determining the free vibration of the patterns. The coefficient of internal friction λ was found in the process of testing the fiberglass for durability.

The ability of ultrasonic waves to travel in a web direction over a minimum time was also established by Golfman.[11]

The velocity of ultrasonic waves propagation was determined as:

$$V_0 = L/t \ 10^3 = L \ f \ 10^3 \tag{5.46}$$

where:
L is the length between the two acoustic heads;
t is the time taken for the ultrasonic oscillations to reach from one head to
the other;
f is the frequency of ultrasonic wave propagation.

In the real dynamic conditions, the internal friction that results from these actions has some delay to ultrasonic wave propagation.

$$V_d = L \ f \ 10^3 \lambda \tag{5.47}$$

The velocity of ultrasonic waves in lattice structures will be:

$$V = V_0 - V_d = L \ f \ 10^3 (1 - \lambda) \tag{5.48}$$

We determined that the modulus of elasticity under angle α as:

$$E\alpha = V^2_\alpha \rho (1 - \mu_{1\alpha}\mu_{2\alpha}) \tag{5.49}$$

The transverse displacement S for a rotor blade in the polar coordinates can be determined as:[9]

$$m \frac{\partial^2 S}{\partial t^2} = F \left(\frac{\partial^2 Sx}{\partial x^2} + \frac{\partial^2 S_R}{\partial R^2} + \frac{\partial^2 S_0}{\partial \theta^2} \right) \tag{5.50}$$

Here:
m is a mass of rotor blade;
F are the outside forces , $F = (L - D - W \sin \theta) \sin \Omega t$;
Sx, S_R, S_θ are the transverse displacements in the polar coordinates.

We can represent the section areas S_x, S_R, S_θ as:

$$S_x = \Upsilon_x\, X^2 : S_R = \Upsilon_R\, R^2; S_\theta = \Upsilon_\theta\, \theta^2$$

We can replace the angle deformation following as:

$$\Upsilon_z = \frac{Mz^R}{E_1 I_z}; \quad \Upsilon_R = \frac{My^R}{E_1 I_R}; \quad \Upsilon_0 = \frac{Mx^R}{G_{xz} T_x}$$

We now substitute the angle deformation and, after differentiation, Equation (5.50) can be shown as:

$$M\frac{\partial^2 S}{\partial t^2} = 2(L - D - W \sin\theta)\sin\Omega t\left(\frac{Mz^R}{E_1 Iz} + \frac{My^R}{E_1 I_R} + \frac{Mx^R}{G_{xz} T_x}\right) \qquad (5.51)$$

We can substitute time t for natural frequency f and, as a result, we can find a material properties structure as well as mass of rotor blade (m), critical damping coefficient (δ), stiffness of rotor blade (Q_{11}), vibration characteristics, forcing frequency (Ω), and natural frequency(f).

The natural frequency f can be found by solving the differential equation for an orthotropic rotor blade in a polar coordinate:

$$\frac{\partial^2 f}{\partial t^2} + \frac{g}{h\eta}\left(D_1\frac{\partial^4 f}{\partial x^2} + 2D_3\frac{\partial^4 f}{\partial x^2 \partial R^2} + D_2\frac{\partial^4 f}{\partial \theta^2}\right) = 0 \qquad (5.52)$$

Here: D_{ij} is the stiffness of the rotor blade from the bending moment.

$$D_1 = Q_{11}S_x \; D_2 = Q_{22}S_{R'} \; D_3 = (Q_{12} + 2Q_{66})S_\theta \qquad (5.53)$$

Stiffness constants Qij are determined only: $Qij = Qji$;
Tx is a geometrical stiffness;
E_1 is the modulus of elasticity in the warp-x direction;
E_2 is the modulus of elasticity in the fill-y direction;
$\mu_{12}\,\mu_{21}$ are the Poisson's ratio.
h is the height of the rotor blade;
g is the density of the fiber;
η is the acceleration due to gravity.

In the case of a free vibration rotor blade, we use the boundary conditions:
If $x = 0$; $x = R$; $f = 0$;

$$\frac{\partial^2 f}{\partial x^2} + \mu_{21} \frac{\partial^2 f}{\partial R^2} = 0$$

If $z = 0$; $z = h$; $f = 0$;

$$\frac{\partial^2 f}{\partial x^2} + \mu_{12} \frac{\partial^2 f}{\partial R^2} = 0 \tag{5.54}$$

h is the height of the section of a propeller blade and R is the radius of a propeller blade.

These boundary conditions are known by the function of deflections:[6]

$$f_{mn} = \sin \frac{m\pi x}{R} \sin \frac{n\pi y}{h} \tag{5.55}$$

Here, m and n are whole digits and are determined as a number of semi-waves in the x and z direction.

$$\text{Wet designate } k = \frac{R}{h};$$

k is the present geometrical parameter (relationship of radius and propeller blade to height).

We can determine natural frequencies f_{mn} as:

$$f_{mn} = \frac{\pi^2}{h^2} \left(\frac{g}{h\eta} \right)^{1/2} \left[D_1 \left(\frac{m}{k} \right)^4 + 2D_3 n^2 \left(\frac{m}{k} \right)^2 + 2D_3 n^4 \right]^{1/2} \tag{5.56}$$

The frequency of the basic tone (m = 1, n = 1) will be:

$$f_{11} = \frac{\pi^2}{h^2} \left(\frac{g}{h\eta} \right)^{1/2} \left(D_1 + 2D_3 k^2 + D_2 k^4 \right)^{1/2} \tag{5.57}$$

The frequency of the second tone (m = 2, n = 2) will be:

$$f_{22} = \frac{4\pi^2}{R^2} \left(\frac{g}{h\eta} \right)^{1/2} \left(D_1 + 2D_3 k^2 + D_2 k^4 \right)^{1/2} \tag{5.58}$$

The frequency of the third tone (m = 3, n = 3) will be:

$$f_{33} = \frac{9\pi^2}{R^2}\left(\frac{g}{h\eta}\right)^{1/2}\left(D_1 + 2D_3k^2 + D_2k^4\right)^{1/2} \tag{5.59}$$

5.5 Experimental Analysis

The main failure mode of actual rotor blades includes bending and torsion moments. In a work by Hou and Gramoll,[12] it was shown that the results of testing of conical lattice structures was not stable. The low failure was due to microbuckling and is commonly referred to as *fiber kinking*. Fiber kinking generally occurs because of a weak matrix, which is due in part to the epoxy not curing completely or a deficiency of the hardening agent during the manufacturing process.

In their investigation, the fiber density that was selected had g = 1770 kg/m³, and the acceleration due to gravity was η = 9.81m/sec², D_{ij} is a stiffness of the rotor blades from the bending moments, and D_1 = 145.2 kg/m² , D_2 = 50.57 kg/m² , D_3 = 120.9 kg/m². All the parameters that were used for natural frequencies were determined.

Small specimen static test results are listed in Table 5.1.

The value of the natural frequencies for the rotor blades depends on the variation of the geometrical parameters k, which are seen in Table 5.2.

The damping coefficient δ_{1111} following Golfman[11] in the warp direction was .01825, δ_{2222} in the fill direction .01933, δ_{1212} in 45° diagonal direction was .02.

Finally, the value of the natural frequencies also was dependent on the variation of the geometrical parameters.

In the case of vibration in the longitudinal direction when the mechanical amplitude was changed from $A_{m1} = A_{e1} = 1$ to $A_{m2} = A_{e2} = -1$, the force frequencies are changed to:

$$\Omega_1 = \frac{E_0}{-\Sigma Rt}; \quad \Omega_2 = \left(\frac{E_0 - \Sigma 1/C}{-\Sigma Lt^2}\right)^{1/2} \tag{5.60}$$

We then input the natural frequencies f = 1/t in Equation (5.34) and get:

$$\Omega_1 = \frac{E_0 f}{-\Sigma R}; \quad \Omega_2 = f\left(\frac{E_0 - \Sigma 1/C}{\Sigma L}\right)^{1/2} \tag{5.61}$$

If the force frequencies change due to variations, then voltage E_0, the inductance (L), resistance (R) and reciprocal of capacitance (1/C) change. The model 352C23ICP accelerometer measures just 28 mm × 5.7 mm and

TABLE 5.1

Small Specimen Static Test Results

	Graphite-Epoxy Prepreg Requirements				Pultrusion (Average of 3 to 5 Specimens)Postcured			
	-65°F	RT	160°F	250°F	-65°F	RT	160°F	250°F
Flexural Strength, ksi	@ (.95RT)	200	(.70RT)	(.65RT)	272	213	130	
Flexural Modulus, psi $\times 10^6$	(.95–1.05RT)	16–18	(.95RT)	(.65RT)	—	17.1	15.8	13.4
Interlaminar Shear Strength, psi	(.95RT)	12,000	(.70RT)	(.65RT)	—	13500	11800	7600
Tensile Strength, Ksi	—	—	—	—	192,0	188,3	189,0	—
Tensile Modulus, psi $\times 10^6$	—	—	—		18,0	17,8	17,9	18,1
Transverse Tensile Strain, in/in	—	4,000	—	—	—	—	—	—

Note: @= Based on 60% fiber volume.

TABLE 5.2

Value of the Natural Frequencies for the Rotor Blades

Radius, m	Height, m h	k =R/h	$f_{11}\times 10^{-4}$	$f_{22}\times 10^{-4}$	$f_{33}\times 10^{-4}$
2	.2	10	5.38	21.52	48.42
4	.2	20	5.28	21.12	47.52
6	.2	30	5.26	21.04	47.34
8	.2	40	5.26	21.04	47.34
10	.2	50	5.23	20.92	47.07

weighs only 0.2 g. The sensor employs a shear-mode, piezoceramic element that generates a 5 mv/g output signal with a frequency response from 2 Hz to 10 kHz. The resonance frequency is specified as greater than 70 kHz.

We varied the voltage from 10 v to 100 v using a decrement of 10; the natural frequency from 10 Hz to 100 Hz; the resistance from 50 to 100 ohms with a 10 decrement; the inductance from .5 mH to 2.5 and the reciprocal of capacitance (1/C) from 1 mF to 2.5. All the terms for the electrical parameters which we used are referred to in Gibilisco.[13]

The principal signal transmission to a propeller or helicopter blade by electrical signal transforms to an ultrasound provide by a piezoceramic transducer. The current registration sensor passes the ultrasound signal to the current conductor elements, which are molded simultaneously into the blade structure. Therefore, we can reduce the vibration frequency acting directly on a structural blade.

5.5.1 Conclusions

1. For composite wind turbine blades, we found a solution to determine the normal and shear stresses in the polar coordinates including the blade profile function.

2. The natural frequencies for the second and third tone in case of storms and hurricanes increase up to four to nine times compared with the natural frequencies for the basic tone and can be put in a favorable area outside the operational rpm range.

3. The circuit device used as a countermeasure is a potential for compensating vibration. This was shown in the use of the reciprocal of capacitance (1/C) and the storage inductance L.

4. Therefore, we can manage force frequencies more flexibly compared to manipulation with only resistance.

5. A methodology was developed to reduce vibration, which is capable of activating an electrical circuit by transferring mechanical energy into electrical energy and has helped designers reduce vibration by 50 percent.

5.6 Mechanical Measurements Deformations in Hybrid Turbine Blades

5.6.1 Introduction

In spite of existing optical and laser systems for measurement deformations, more reliable mechanical measurement deformations are used for fiberglass turbine blades.

5.6.2 Strain–Stress Relation

Hook's law is the linear strain–stress relation for fiberglass/carbon power turbine blades. It is derived from the elastic energy as a basic postulate in the theory of elasticity. If the axis of geometry blades is consistent with warp direction of fiber, Hook's law is represented by Equation (5.62):[14]

$$e_x = \frac{\sigma_x}{E_x} - \frac{\mu_{xy}}{E_y}\sigma_y - \frac{\mu_{zx}}{E_z}\sigma_z$$

$$e_y = \frac{\mu_{xy}}{E_x}\sigma_x - \frac{\sigma_y}{E_y} - \frac{\mu_{zy}}{E_z}\sigma_z$$

$$e_z = \frac{\mu_{xz}}{E_x}\sigma_x - \frac{\mu_{yz}}{E_y}\sigma_y - \frac{\sigma_z}{E_z}$$

$$\gamma_{xy} = \frac{\tau_{xy}}{G_{xy}};$$

$$\gamma_{yz} = \frac{\tau_{yz}}{G_{yz}};$$

$$\gamma_{zx} = \frac{\tau_{zx}}{G_{zx}}$$

(5.62)

Here:

e_x, e_y, e_z are linear deformations in x, y, z directions;

γ_{xy}, γ_{yz}, γ_{zx} are angle deformations;

σ_x, σ_y, σ_z are the tension/compression stresses in x, y, z directions;

τ_{xy}, τ_{yz}, τ_{zx} are the shear stresses act in plain xy and transverse xz, yz.

E_x, E_y, E_z are the modulus of elasticity in x, y, z directions;

G_{xy}, G_{yz}, G_{zx} are the shear modulus of elasticity;

μ_{xy}, μ_{xz}, μ_{yz} are the Poison's ratio, the first index indicate stress direction, the second index indicate strain direction.

Investigation of strength hybrid turbine blades begin from measuring deformation and estimating stresses in different points of the blades. The widely used resistance strain gauges, which are the most universal, have recently become popular for measuring elastic deformations in longitudinal, transverse, and diagonal 45° directions.[15]

The gauges loop's influence on accuracy measurement deformations can be corrected using coefficient χ:[16]

$$\chi = \frac{k_x/k_y + \mu_{xy}}{1 + \mu_{xy}k_x/k_y}$$

(5.63)

Here: k_x and k_y = sensitivity coefficients in x, y directions was determined by calibrated fiberglass specimens. Normal and shear stresses acting in root sections will be determined with influence coefficient χ Equation (5.63).

$$\sigma_x = A\,E_x\left[\varepsilon_x\left(1 - \mu_{xy}\chi - \varepsilon_y(\chi - \mu_{yx})\right)\right]$$

$$\sigma_y = A\,E_y\left[\varepsilon_y\left(1 - \mu_{yx}\chi - \varepsilon_x(\chi - \mu_{xy})\right)\right]$$

(5.64)

$$\tau_{xy} = G_{xy}\left[2\varepsilon^{45} - \frac{1 - \mu_{xy}\chi}{1 + \chi}\left(\varepsilon_x + \varepsilon_y\right)\right]$$

$$\text{Coefficient } A = \frac{1 - \mu_{xy}\chi}{\left(1 - \mu_{xy}\mu_{yx}\right)\left(1 - \chi^2\right)}$$

(5.65)

Modulus of elasticity E_x, E_y, G_{xy}, and Poisson ratios μ_{xy} and μ_{yx} was predetermined on the fiberglass specimens; deformations ε_x, ε_y, and ε_{45} was measured when load reached the beginning of failure.

A power wind turbine blade with resistance strain gauges is shown in Figure 5.2.

Stresses for wind turbine blades in x, y directions (see Figure 5.2) can be calculated using Equation (5.64).

Here: μ_{xy}, μ_{yx} coefficients of Poisson's ratio

ρ is density of hybrid materials (80 percent fiberglass and 20 percent carbon fiber);

ω is a velocity of rotation blade (22–25 m/s, see Chapter 2, rotor design);

K is a coefficient of profile.

Following Golfman,[17] coefficient of profile has been found as:

$$K = \frac{27}{16}\frac{e_{k^2}\lambda}{(\alpha - \beta)}$$

(5.66)

Here: e_k is a maximum thickness, λ is coefficient of anisotropy, relations of longitudinal stiffness to transverse stiffness;

$$\lambda = \frac{E_{xy} + G_{xy}}{E_{xz} + G_{xz}}$$

(5.67)

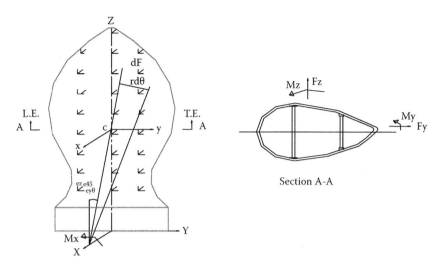

Power turbine blade with resistance strain gauges

FIGURE 5.2

A power wind turbine blade with resistance strain gauges. (From Bronsted, Lithoit, and Lystrup. 2005. *Composite materials for wind power turbine blades.* Copenhagen: Technical University of Denmark, Riso National Laboratory for Sustainable Energy. With permission.)

where:

E_{xy}, E_{xz} are the modulus of normal elasticity in x, z directions;

G_{xy}, G_{xz} are the modulus of shear elasticity in x, z directions.

α is a distance from end of profile to center;

β is a distance from center to end of profile.

5.7 Mechanical and Thermal Properties

5.7.1 Introduction

One of the most important characteristics of all prepregs is the way the material behaves when acting with mechanical and thermal loads. With temperature changes from -20°C to +40°C, this mechanical and thermal data is critical to almost every application including power turbine blades. Although an accurate determination as to whether it is suitable for use at a particular temperature can only be made by testing the product in the loading configuration that it will see in service, the data presented here can be used as a guide for initial product selection.

5.7.2 History of Investigating Mechanical Properties

The wind turbine blades are designed in relation to stiffness and fatigue. Ratio of high stiffness and high strength to weight are two characteristics that depend on selection of the carbon/glass relationship. Usually, the relation of glass/carbon with matrix resin is: 70 percent reinforced carbon, 30 percent resin. This relationship has a large influence on fatigue strength. Basic properties of individual constituencies include the stiffness predictions of fabric and matrix use. The calculations are experimentally verified by static testing that measures tension, compression, and shear stiffness and stress. All values are used in strength criteria for estimated lifespan predictions. Stiffness, static strength, and fatigue performance are measured by testing test coupons or components.

Uniform static tests are carried out under quasi-static loading, and fatigue tests done under varying loading conditions. Properties are measured under tensile loads, compression loads, shear loads, or combinations of this in bi- and multiaxial loading modes. During the 1980s and 1990s, three European projects, from 1986 to 1996,[18-20] focused on research on the mechanical behavior of the wind turbine blades and the materials. In this program, they measured the static and fatigue properties of glass–polyester and glass–epoxy materials. Techniques were developed with design curves based on constant load amplitude tests and on variable amplitude tests using stochastic load histories being established. The results from these projects were collected in two books edited by C. W. Kensche[21] and S. R. M. Mayer.[22] E. C. Smet and P. W. Bach collected the data into a database FACT.[23] Supported by Sandia National Laboratories, Montana State University (MSU) has worked intensively since 1989 on characteristics of composite materials for wind turbines. The result from this nationally supported program was collected in a large DOE (Department of Energy)/MSU database that was first presented in 1997 by J. F. Mandrell and D. D. Samborsky.[24] Since 2001, the database has been updated annually.[25]

The European activities have continued in a currently running European project (2001–2006). OPTIMAT,[26-28] under the Fifth European Union (EU) framework program, provide accurate design recommendations for the optimized use of materials within wind turbine rotorblades and to achieve improved reliability. The project is investigating the structural behavior of the composite material (glass–epoxy) under the unique combinations experienced by rotorblades, such as variable amplitude loading, complex multiaxial stress states,[30-31] extreme environmental conditions, and thick laminates and their possible interactions. Techniques will be developed for damage mechanisms,[29] life extension, condition assessments and repair.

5.7.3 Testing Mechanical and Thermal Properties of the Prepreg Laminates

We tested the key mechanical properties of the laminated plates (25 × 2.4 mm) of the different lengths including tensile and compression strength (modulus) at 0, ±45°, and 90° configurations; shear strength/modulus, and interlaminar shear strength, as described by Y. Golfman.[17] Thermal properties of interest, coefficient of thermal expansion (CTE), and thermal conductivity, were examined. The combined test panel was based on the following ASTM (American Society for Testing and Materials) standards:

- ASTM D638 (ref. D3039/D3039M) Test Method for Tensile Properties of Polymer Matrix Composite Materials
- ASTM D696 (ref. D3410) Test Method for Compression Properties of Polymer Matrix Composite Materials
- ASTM D732 Shear Strength of Plastics by Punch Tool
- ASTM D903 Peel or Stripping Strength of Adhesive Bonds
- ASTM D696 Coefficient of Linear Thermal Expansion of Plastics

We also examined the microscopy of the laminate surface after thermal cycling for presence of microcracking.

Mechanical properties of prepreg laminates, consisting of traditional carbon fibers and epoxy matrix, can be modified using carbon nanotubes. Basic mechanical properties of unidirectional laminates (stiffness, strength, fracture toughness, etc.) are summarized. Next, quasi-isotropic laminates are subjected to tension, compression, flexural, and compression after impact (CAI) tests.

Figure 5.3 demonstrates the tensile test setup with measurement of longitudinal and transverse strains by video-extensometry. And, advanced fixture for combined compression load transfer is represented in Figure 5.4.

5.7.4 Conclusions

1. The increasing demands for design toward stiffness and fatigue strength have required development of more accurate and reliable testing techniques.
2. Improvement of stiffness and strength and with no adverse effects on mechanical properties due to carbon nanotubes dispersion are experimentally verified.

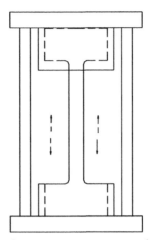

Tensile test set-up in instron machine
with measurements of longitudinal and
transverse strains.

FIGURE 5.3
Tensile test setup with measurement of longitudinal and transverse strains by video exten-
sometry. (From Bronsted, Lithoit, and Lystrup. 2005. *Composite materials for wind power turbine blades.* Copenhagen: Technical University of Denmark, Riso National Laboratory for
Sustainable Energy. With permission.)

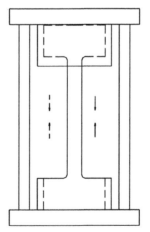

Compression test set-up in instron machine
with measurements of longitudinal and
transverse strains.

FIGURE 5.4
Advanced fixture for combined compression load transfer. (From Lethe, De Cuper et al. 2009.
Journal of Mechanical Science and Technology 23 (4): 4. With permission.)

5.8 Fatigue Strength And Weibull Analysis

5.8.1 Introduction

The application of hybrid anisotropic composites, such as reinforced fiber-glass with carbon/epoxy materials, in large wind turbine blades will be critical for a source of alternative energy.

Wind turbine blades are designed to include bending and torsion moments, aerodynamic wind cutting forces, and, during severe loading, high numbers of fatigue cycles and reverse tension-compression loads. The major static strength and stiffness properties depend primarily on fiber type, content, and orientation.

The fatigue of composite laminates appropriate for wind turbine blades has been a topic of research studies.

5.8.2 Fatigue Strength Prediction

The purpose of this research is the prediction of fatigue strength of composite wind turbine blades under long-term service, which is dependent on several technological factors.

In the large turbine blades fabricated from composites, i.e., fiberglass or carbon/epoxy material, there is a significant lower strength than on samples. Some authors call this *scaling effects* and this strength construction is shown as *technological defects*.[32]

$$\sigma = \sigma_0 / N_i^{\frac{1}{2}} \qquad (5.68)$$

where:

σ_0 is a sample strength;
N_i is the number of layers.

Strength reduction can be explained as a result of the influence of technological factors in resin formulation and curing, and fiber distortion. We consider these technological factors as the defects of structural laminates and the technological behavior as shrinkage and warpage, which results from thermal stress during the molding process.

Bailey, Curtis, and Parvizi[33] have shown, using a simple equilibrium model, that the thermal residual stress transverse to the fibers in a constrained 90° ply can be expressed by Equation (5.69):

$$\sigma_{th} = \frac{\Delta T t^c E_2 E_c \left(\alpha^c - \alpha_2 \right)}{E_2 t_2 + E^c t^c} \qquad (5.69)$$

where:

ΔT is the change in temperature and t^c, α^c, and E^c are the thickness, thermal coefficient of expansion, and stiffness respectively of the constraining plies;

t_2, α_2, E_2 are the thickness, thermal coefficient of expansion, and stiffness, respectively, of the 90° piles.

This stress is introduced upon cool down from the curing temperature due to the mismatch in the coefficient of thermal expansion of the adjacent piles in a laminate.

From the prediction strength of every layer, we can approximate the average strength of all construction and answer the question of how long this construction will be serviceable.[34]

We consider that every layer of construction has strong orthotropic properties, and the construction has a homogeneous structure and is equally impregnated by epoxy or other isotropic resin. Optimal structure has oriented fibers whose direction of reinforcement coincides with the direction of acting normal stresses along the axis x, y, z. (Figure 5.5).

Fatigue strength can be predicted as a linear correlation between compression strength and the function Φ(σ), see Equation (5.70).

$$\sigma_{-1}=\sigma_s\Phi(\sigma) \qquad\qquad (5.70)$$

where:

σ_{-1} is fatigue strength in x, y, z direction.

σ_s is compression strength in x, y, z direction.

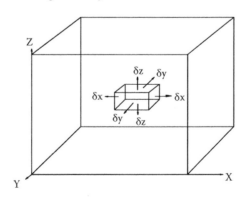

Fiber orientation in orthotropic composites.
Stress components coincide with fiber orientation X,Y,Z.

FIGURE 5.5

Fiber orientation in orthotropic composites. Stress components coincide with fiber orientation x, y, z. (From Lethe, De Cuper et al. 2009. *Journal of Mechanical Science and Technology* 23 (4): 4. With permission.)

Function $\Phi(\sigma)$ can be shown as the Weibull distribution function, see Equation (5.71):

$$\Phi(\sigma) = 1 - P(t) \qquad (5.71)$$

Here: $P(t)$ is the probability of collapse in the local part of the construction from compression strength.

We assume that the general strength equals unity and $P(t)$ has been subordinated to the normal distribution law, see Equation (5.72):

$$P(t) = \frac{1 - t \frac{2}{2}}{(2p)^{1/2}} \qquad (5.72)$$

where:

parameter t is equal to:

$$t = \frac{\sigma_{bi} - \sigma_{bm}}{Sj} \qquad (5.73)$$

σ_{bi} is a current strength in x, y, z direction;

σ_{bm} is a middle strength in x, y, z direction;

Sj is a sample standard deviation for each environment, see Equation (5.74):

$$S_j^2 = \frac{1}{n_J - 1} \sum_{I=1}^{n_j} \left(\sigma_{bi} - \sigma_{bm} \right)^2 \qquad (5.74)$$

Here: n_j is the number of testing samples.

Calculating the sample mean σ_{bm}, see Equation (5.75):

$$\sigma_{bm} = \frac{1}{n_i} \sum_{n=1}^{n_j} \sigma_{bi} \qquad (5.75)$$

For a single test condition (such as $0°$ compression strength), the data were collected for each environment being tested. The number of observations in each environmental condition was n_j where the j subscript represents the total number of environments being pooled. If the assumption of normality was significantly violated, the other statistical model should be investigated to fit the data. In general, the Weibull distribution provides the most conservative basic value.

In R. Talreja,[34] the Weibull distribution function is given in Equation (5.76):

$$\Phi(X,A,B,C,) = 1 - \exp\left\{-\left(\frac{X-A}{B}\right)^C\right\} \qquad (5.76)$$

Here, parameters are:
  $X > 0; B > 0; C > 0;$
  X, A, B, C each are equal to discrete symbols.

For strength distribution, we designate:
  $X = \sigma_{bi};$
  $A = \sigma_{bm};$
  $B = Sj;$
  $C = N;$
  N-base of testing.

Therefore, Equation (5.76) for fatigue strength represents Equation (5.77):

$$\Phi(\sigma_{bi}, \sigma_{bm}, Sj, N) = 1 - \exp\left\{-\left(\frac{\sigma_{bi} - \sigma_{bm}}{Sj}\right)^N\right\} \qquad (5.77)$$

If we consider that:

$$\Phi(\sigma_{bi}, \sigma_{bm}, Sj, N) = 1 - P(t) = \exp\left\{-\left(\frac{\sigma_{bi} - \sigma_b}{Sj}\right)^N\right\} \qquad (5.78)$$

We get the logarithmic equation, see Equation (5.79):

$$\ln[1 - P(t)] = N \ln(\sigma_{bm} - \sigma_{bi}) - N \ln Sj \qquad (5.79)$$

A straight line in logarithmic coordinates shows in Equation (5.79). Base of testing N can determine the inclination of this straight line. Therefore, the period of testing N shall be determined in Equation (5.80):

$$N = \frac{\ln[1 - P(t)]}{\ln(\sigma_{bm} - \sigma_{bi}) - \ln Sj} \qquad (5.80)$$

If we propose that the shear stress responsible for delaminating of the composite, the period of testing N shall be determined as:

$$N_1 = \frac{1n\left[1 - P(t_1)\right]}{Ln\left(\tau_{bm} - \tau_{bi}\right) - 1nSj}$$ (5.81)

Here:

$P(t_1)$ is the probability structure collapse responsible from interlaminar shear stress;

τ_{bi} is the current shear stress acting in interlaminar layers;

τ_{bm} is the middle significant shear stress;

$S\tau$ is the middle square deviation.

Middle square deviation can be found using Equation (5.82):

$$S\tau^2 = \frac{1}{n-1}\left(\tau_{bi} - \tau_{bm}\right)^2$$ (5.82)

Shear strength samples molded by fiberglass, carbon epoxy, or hybrid fiberglass/carbon-tested use two methods: quad lap shear test (peel test), and tear test.

Shear strength fiber compatibility test is shown in Figure 5.6.

If the interlaminar shear strength is responsible for collapse construction, fatigue strength can be predicted as:

$$\tau_{-1} = \tau_s \Phi(\tau)$$ (5.83)

Function $\Phi(\tau)$ is shown as:

$$\Phi(\tau) = 1 - P(t_1)$$ (5.84)

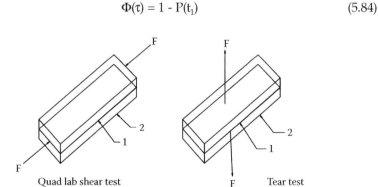

Quad lab shear test

1-First glass fiber layer
2-Second carbon fiber layer
shear strength test

Tear test

FIGURE 5.6
Shear strength test: (1) first glass fiber layer; (2) second carbon fiber layer.

Here: $P(t_1)$ is the probability of collapse of the local part of construction from the interlaminar shear strength.

$$\text{Parameter } \tau_1 = \frac{\tau_{bi} - \tau_{bm}}{S\tau} \tag{5.85}$$

The Weibull distribution function is shown as the example of the Laplace function. The Laplacian transformation is a powerful method for solving linear differential equations arising in engineering mathematics.[35] Thus, we designate:

$$P(t) = L(t) = \int_0 e^{-bt} f(t) dt \tag{5.86}$$

The function $L(t)$ is called the Laplacian transform of the original function $f(t)$. Furthermore, the original function $f(t)$ is called the inverse transform function and will be denoted by $L^{-1}(F)$, that is:

$$f(t) = L^{-1}(F) \tag{5.87}$$

$$L(e) = \frac{1t - t^{2/2}}{(2p)^{1/2}} \int_0 e \tag{5.88}$$

Function $L(e)$ is a two parameter function that represents the law of normal distribution function and the prediction probability with variation strength parameters σ_{bi}.

5.8.3　Static and Dynamic Fatigue Strength

Reinforced fiberglass or carbon/epoxy are very sensitive to the equal static load repeatedly applied in a long period of time, which is why it is called *static fatigue of the material.*

The limit of fatigue for graphite/epoxy on the base of 1,000 cycles equals 0.6 to 0.7 from the limit of the static strength. Bending does not increase due to beneficial elastic properties. Loops of hysteresis from loading and unloading are practically identical. Residue of deformation accumulated after testing samples for 1,000 cycles have no significant value. Dynamic testing has the support significance of internal heat.

Internal heat depends on harmonic frequencies and the amplitude of vibration, and the heat of the composites is connected with hysteresis losses. The quantity of heat is increased when there is an increase in frequency of vibration. If the frequency is 1,000 cycles per minute, the internal heat of the sample will be 50 to 70°C.

If the frequency decreases to 300 cycles per minute, the internal heat simultaneously is reduced to 25 to 30°C. The very important characteristic of dynamic properties is the internal dispersion of energy (critical damping coefficient).[36] During the harmonic vibration, which can reach 1×10^6 cycles, there is no visible cracking. It does not mean that there is no internal cracking.

It is very important to use nondestructive ultrasonic evaluation for the determination of the dynamic modulus of elasticity,[37] and not visible cracking propagation.

The criteria of quality for dynamic fatigue after 1×10^6 to 5×10^6 cycles have shown crack visibility. However, the samples with visibility cracks after testing have very high strength characteristics. Therefore, the value of predicting fatigue strength is somewhat lower than test results.

Correlation between dynamic fatigue and the number of cycles in the case of harmonic bending in the graphite/epoxy composites exist.

5.8.4 Experimental Investigation

Researchers did between 5 to 20 tests on small specimens generating the fatigue strength.[35] The diagram describing the relations between loading and number of cycles has a name: Wohler diagram (Figure 5.7)

Table 5.3 represents the parameters of the compression strength for graphite/epoxy composites.[36] We tested six samples and determined the sample standard deviations.

Also, we determined three parameters: t, probability P(t) using tabulating function (Table 5.4), and function $\Phi(\sigma)$ following Equation (5.71).

Fatigue strength prediction has a good correlated with compression strength (Figure 5.8).

Table 5.5 represents the parameters of interlaminar shear strength for graphite/epoxy composites. We tested six samples and determined the sample standard deviation.

Also, we determined three parameters: t_1, probability $P(t_1)$ using tabulating function (Table 5.4), and function $\Phi(\tau)$ following Equation (5.84).

Fatigue strength prediction using Laplacian function Equation (5.88) is represented in Table 5.6.

Tabulating Laplacian functions are given in Table 5.7.

The number of samples N, the sample mean σ_{bm}, τ_{bm}, and sample standard deviation Sj and St was selected from the pooled data. The normal distribution B-basis and A-basis allowable was calculated using the pooled mean, standard deviation, and tolerance factors for each environment.

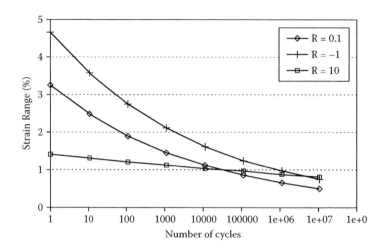

FIGURE 5.7

The Wohler diagram showing 50 percent regression lines based on data from the OPTIMAL database.

TABLE 5.3

Fatigue Compression Strength Prediction for Graphite/Epoxy Composites

$\sigma_s \cdot 10^{-3}$ MPa/psi	$\sigma_m \cdot 10^{-3}$ MPa/psi	$S_j \cdot 10^{-3}$ MPa/psi	Parameter t	P(t)	$\Phi(t)$ $\sigma_{-1} \cdot 10^{-3}$	Fatigue Strength MPa/psi
.37/54			.412	.3668	.6332	.24/34.2
.36/52			.24	.3885	.6115	.22/31.8
.34/50	.34/49.16	.024/3.46	.07	.3980	.6000	.20/30.0
.33/48			.009	.3989	.6010	.198/28.8
.32/46			.025	.3876	.6124	.19/28.17
.31/45			.035	.3765	.6235	.19/28.05

TABLE 5.4

Density of Probability Function (First Approach)

t	0	1	2	3	4	5	6	7	8	9
0,0	3989-4	3989	3989	3988	3986	3985	3998	3980	3977	3973
0,1	3970-4	3965	3961	3956	3951	3945	3939	3932	3925	3918
0,2	3910-4	3902	3894	3885	3876	3867	3857	3847	3836	3825
0,3	3814-4	3802	3790	3778	3765	3752	3739	3726	3711	3697
0,4	3683-4	3668	3653	3637	3621	3605	3589	3572	3555	3538
0,5	3521-4	3503	3485	3467	3448	3429	3410	3391	3372	3352
0,6	3332-4	3312	3292	3271	3251	3230	3209	3187	3166	3144

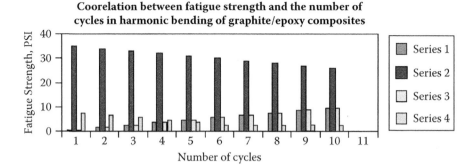

FIGURE 5.8
Fatigue strength prediction correlations with compression strength.

TABLE 5.5

Fatigue Interlaminar Shear Strength Prediction for Graphite/Epoxy Composites
(First Approach)

$\tau_s \cdot 10^{-3}$ MPa/ psi	$\tau_s \cdot 10^{-3}$ MPa/ psi	$S\tau^* 10^{-3}$ MPa/ psi	t_1	$P(t_1)$	$\Phi(t_1)$	$\tau_{-1} \cdot 10^{-3}$ MPa/ psi
.07/10.4			.692	.3144	.685	7.12
.07/10.2			.384	.3711	.628	6.41
.068/10	.068/9.95	.004/.65	.076	.3980	.602	6.02
.067/9.8			.230	.3885	.611	5.99
.066/9.7			.384	.3711	.628	6.09
.066/9.6			.538	.3448	.655	6.28

5.8.5   Concluding Remarks

1. A methodology for the prediction of the fatigue strength of anisotropic materials like fiberglass, carbon/epoxy, and hybrid matrix composites was developed.

2. Probability of local cracking can be predicted using mathematical functions; the first and second approach the law of normal distribution.

3. In spite of the visibility of cracks, the samples have a general strength. However, it is very important to use nondestructive evaluation methods for the determination of cracks that are not visible.

TABLE 5.6

Fatigue Compression and Interlaminar Strength Prediction for Graphite/Epoxy Composites (Second Approach)

$\sigma_s \cdot 10^{-3}$ MPa/ psi	$\sigma_m \cdot 10^{-3}$ MPa/ psi	$S_j \cdot 10^{-3}$ MPa/ psi	L(t) Parameter t	Φ(t) (equation (5.71))	Fatigue Strength MPa/ psi	$\sigma_{-1} \cdot 10^{-3}$
.34/50	.34/49.16	.023/3.46	.070	.279	.721	.29/36.05

$\tau_s \cdot 10^{-3}$ MPa/ ps	$\tau_s \cdot 10^{-3}$ MPa/ psi	$S\tau^* \cdot 10^{-3}$ MPa/ psi	t_1 MPa/ psi	$L(t_1)$	$\Phi(t_1)$	$\tau_{-1} \cdot 10^{-3}$
.068/10	.068/9.95	.004/.65	.076	.299	.701	.048/7.01

TABLE 5.7

Density of Probability Laplacian Function (Second Approach Equation (5.88))

t	0	1	2	3	4	5	6	7	8	9
0,0	0,0 000	040	080	120	160	199	239	279	319	359
0,1	398	438	478	517	557	596	636	675	714	753
0,2	793	832	871	910	948	987	.026	.064	.103	.141
0,3	792	217	255	293	331	368	406	443	480	517
0,4	554	591	628	664	700	736	772	808	844	879
0,5	915	950	985	.019	.054	.088	.123	.157	.190	.224
0,6	257	291	324	357	389	422	454	486	517	549

5.9 Dynamic Analysis: Fourier Function for Prediction of Fatigue Lifecycle Test

5.9.1 Introduction

In section 5.8, we found the optimum parameters of fiberglass/carbon using the Weibull analysis. This section (5.9) describes how Fourier function can predict fatigue life.

Light weight and high strength anisotropic composites, such as carbon/ carbon, graphite epoxy, fiberglass, etc., under different combinations of applied stress components (biaxial and triaxial stress conditions) pose a challenge to designers for establishing reliable failure criterion. Stimulated stress components in a composite turbine disk under the influence of centrifugal loads and temperature are a typical example. In this research, we propose to use Hook's law relationship between strain and stress components including effect of thermoelasticity called the Duhamel–Neumann law. Our concept was based on determining the stress parameters using equations from Hook's law and Duhamel–Neumann law and find results from positive and negative values dependant on changing temperature and coefficient of thermal expansions for composite wind turbine blades.

5.9.2 Theoretical Investigation

When properties of an orthotropic turbine blade have been changed under acting loads and temperature, the stress–strain relations must be signed in matrix form.[38]

$$\sigma_{ij} = Q_{ij}^{*}(\varepsilon_j - \alpha_j^{*}T) \ (i,j = 1,2,6) \tag{5.89}$$

where:
Q_{ij} are known as the stiffness constants;
α_j are coefficients of temperature expansion;
T is an environmental temperature.

For orthotropic materials, we have 10 stiffness constants (see Equation (5.89)). This relation V. Z. Parton[38] called the Duhamel–Neumann Law. Here:

E_1, E_2 are modules of normal elasticity in warp and fill directions;
μ_{12} ,μ_{21}, μ_{23} ,μ_{32} are Poisson's ratio of material.

The first symbol designates the direction of force and second symbol designates the direction of transverse deformation. α_1, α_2 are coefficients of thermal expansion in longitudinal (warp) and axial (fill) directions.

Summarizing Equation (5.89) for laminate package, we get in longitudinal direction x:

$$\Sigma\sigma_{-1x} = \Sigma Q_x \varepsilon_{xy} - \Sigma Q_x \alpha_x^{*}T) \tag{5.90}$$

and in axial direction y:

$$\Sigma\sigma_{-1y} = \Sigma Q_y \varepsilon_{yx} - \Sigma Q_y \alpha_y^{*}T) \tag{5.91}$$

Here:

$\Sigma\sigma_{-1x}$, $\Sigma\sigma_{-1y}$	= Fatigue stress carbon/fiberglass sandwich;
ΣQ_x, Q_y	= Stiffness of combination carbon/fiberglass in x, y directions;
ε_{xy}, ε_{yx}	= deformations in x, y directions;
α_x, α_y	= coefficients of thermal expansion in x,y directions. This coefficients has anisotropic character;
T	= Temperature gradient

Harmonic fatigue stress is illustrated in Figure 5.9
The Fourier series is described as a function.[39]

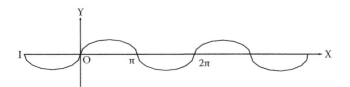

Harmonic Fourier function

FIGURE 5.9
Harmonic Fourier function.

$$Y = 8/\pi(\sin X + 1/3^3\sin3X + 1/5^3\sin5X + \ldots\ldots) \qquad (5.92)$$

$Y = \Sigma\sigma_{-1x,}$ or $\Sigma\sigma_{-1y}$; $X = F_{ND}$; F_{ND} is Duhamel–Neumann function:

$$F_{NDX} = \Sigma Q_x \varepsilon_{xy} - \Sigma Q_x \alpha_x{}^*T \; ; F_{NDY} = \Sigma\sigma_{-1y} = \Sigma Q_y \varepsilon_{yx} - \Sigma Q_y \alpha_y{}^*T \qquad (5.93)$$

$$\Sigma\sigma_{-1x} = 8/\pi(\sin F_{NDX} + 1/3^3\sin3\ F_{NDX} + 1/5^3\sin5\ F_{NDX} + \ldots\ldots) \qquad (5.94)$$
$$\Sigma\sigma_{-1y} = 8/\pi(\sin F_{NDY} + 1/3^3\sin3\ F_{NDY} + 1/5^3\sin5\ F_{NDY} + \ldots\ldots)$$

5.9.3 Experimental Investigation

Prepreg based on reinforced IM (intermediate modulus) fiber, tape T40/800B impregnated with tough epoxy resin Cycom® 5320-1 in tension direction has strength equal to 2,565 MPa. Testing conditions are: -100°F(-73°C) (see Table 3.5). We can see anisotropic character properties strength in 0° tension in 25 to 30 times higher than in transverse direction 90 degree (see Table 3.5).

If sine of Duhamel–Neumann function F_{ND} equals strength σ_{-1} for one cycle:

$$\sin F_{NDX} = \sigma_{-1}$$

In Equation (5.94), we can sign as:

$$\Sigma\sigma_{-1x} = 8/\pi(\ \sigma_{-1} + 1/3^2\,\sigma_{-1} + 1/5^2\sigma_{-1} + \ldots\ldots) \qquad (5.95)$$
$$\Sigma\sigma_{-1y} = 8/\pi(\ \sigma_{-1} + 1/3^2\sigma_{-1} + 1/5^2\sigma_{-1} + \ldots\ldots)$$

Tension test in 0° direction IM Fiber reinforced tape T40/800B impregnate by Cycom 5320-1 shows strength drop (Figure 5.10).

In the horizontal, we indicate number of cycles; in vertical, strength drop in tension test. Compression test in 0° direction, IM fiber reinforced tape T40/800B impregnate by Cycom 5320-1 shows a drop in strength (Figure 5.11).

Shear strength test IM fiber reinforced tape shows also a drop in strength (Figure 5.12).

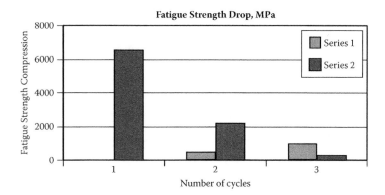

FIGURE 5.10
Tension test in 0° direction IM Fiber reinforced tape T40/800B impregnate by Cycom® 5320-1.

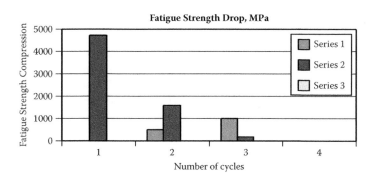

FIGURE 5.11
Compression test in 0° direction, IM Fiber reinforced tape T40/800B impregnate with Cycom 5320-1.

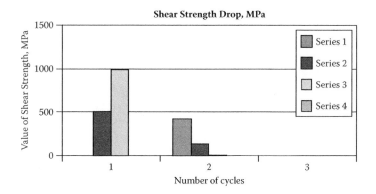

FIGURE 5.12
Shear strength test IM fiber reinforced tape T40/800B impregnate with Cycom 5320-1.

5.9.4   Conclusion

Fourier analysis helps researchers investigate behavior of glass/carbon turbine blades.

5.10   Simulating Dynamics, Durability, and Noise Emission of Wind Turbines

5.10.1   Introduction

In many countries, governments increase the share of renewable power generation through ecologic targets and resolute choices for "green" energy that is clean, indigenous, and inexhaustible. Wind energy is predicted to meet approximately 25 percent of Europe's power demand in the year 2030 and wind turbine markets are also growing fast in the United States and in Asia.

The drive train forms the very heart of a wind turbine. The wind power is converted via the blades into mechanical power on a slow-speed shaft. This power is scaled via a very large gearbox to a high-speed shaft and, finally, transformed via a generator into electrical power. The wind turbine control system together with the power converter guarantees a clean and steady output voltage at constant frequency, irrespective of varying wind conditions.

5.10.2   Engineering Challenges

To obtain certification for a wind turbine, manufacturers have to ensure full system reliability under real-life operating conditions. Noise emissions must remain within prescribed tolerances. Moreover, durability must be assessed to provide a 20-year lifetime with small operation and maintenance costs. Overcoming these challenges involves extensive engineering efforts from the initial concept designs up until the final wind turbine validation and certification. Since extensive tests on full-scale wind turbines are extremely expensive and often dangerous to conduct, manufacturers heavily rely on simulation throughout the development process.

To optimize the design of the gearbox, manufacturers make use of simulation tools to predict the torques that the different shafts need to transmit along the drive train from the blades to the generator. This can be done using simplified codes, which only account for one torsional degree-of-freedom. However, detailed 3D, multibody simulation allows for more in-depth studies, capturing the dynamic behavior of the overall system and its components. For example, the engineering of bearings and gear contacts provides a true challenge to gearbox manufacturers and clearly illustrates the advantages of 3D multibody simulations. Bearings have to endure very high loads and, therefore, are critical in the reliability of the complete system. In-service misalignment of the shafts—as small as a few thousandths of a degree—caused by the compliance of those bearings, influences the gear contact forces and cause unwanted wear on the gear teeth. In order to avoid this wear, gearbox manufacturers can optimize the gear geometry (lead crown and involute crown) based on virtual simulation.[40]

5.10.3   An Integrated Simulation Process

The LMS® Virtual.Lab Motion (Leuven, Belgium), wind turbine engineers can accurately model contacts between gearbox components. This allows them to efficiently predict the transmission of loads between components like gears, shafts, or bearings, while taking the flexibility of these components into account.

Based on the wind load cases, LMS Virtual.Lab evaluates the stresses on each component and predicts their structural static reliability. Using the same Virtual.Lab simulation environment, users can perform durability, vibration, and noise analyses in a straightforward way . In addition, the seamless integration and the preserved associativity within LMS Virtual.Lab avoid time-consuming and error-prone transfer of data and allow development team to perform fast optimization loops.

5.10.4   Multibody Simulation to Assess Dynamic Behavior

Multibody simulation is used to assess the structural reliability of the gearbox and to make sure it resists the extreme and unpredictable loads from the wind and doesn't break under high or concentrated stresses. In a standard 1.5 MW wind turbine, the huge input torque of around 1,000 kNm coming from the blades rotating at 15 rpm has to be transferred to realize a gear ratio between the input and output shaft of more than 100 in order to match the rotational speed needed to generate electricity from the generator. Typically, this is done through a three-stage gearbox design; a first stage with planetary gears (offering a large gear ratio and low weight of the gears) and two subsequent parallel gear stages. Next to the gears themselves, the bearings have to be modeled in detail because of the very high loads they support and their key impact on the overall reliability of the wind turbine.

In the first step of the process, the CAD (computer-aided design) geometry is imported or created within LMS Virtual.Lab Motion using the embedded 3D modeling capabilities. The model includes the gears, shafts, bearings, and housing. In order to switch from a pure kinematic analysis to a dynamic analysis, the gearbox model is then extended with flexible bodies (with inertial parameters), connections between them (joints, constraints, forces) as well as controls. At this level, a dynamic simulation is computed and the results can be visualized through 3D animations or 2D graphs from any variable in the model. Various alternatives of the design (from CAD or dynamic parameters) are compared in order to optimize the system with regards to any specific performance attribute. This allows an in-depth understanding of the root causes of its behavior and enables engineers to minimize the risk of failure during subsequent assessment test on prototypes.

LMS Virtual.Lab provides several methods to model the meshing of gears. The most suited is the so-called *gear contact force* element. It is applicable to any kind of gear system that is spur or helical, external or internal. The method also applies to planetary gears that are often used in wind turbines. In case of the gear contact force, the contact formulation is not computed based on one-tone tooth geometry interpenetration. Instead, it is directly derived from the global gear theory for more efficient analytical solving and shorter calculation times. Moreover, it takes into account the variability of the stiffness, which is due to the profile of one single tooth and to the instantaneous number of meshing teeth, according to the Cai and ISO formulations.[41-44]

The equivalent meshing stiffness between the two gears is the sum of the contact stiffness over the number of contacting teeth, which varies in time according to the contact ratio.

Each single tooth equivalent stiffness is also varying during the meshing time, since the bending is bigger at the top than at the root of the tooth. An example of the meshing stiffness variation is shown in Figure 5.13, assuming that the contact ratio is 2.5, meaning the number of contacting.

Teeth vary from two to three. The total contact stiffness, consequently, has a fluctuation nature (oscillation around a static stiffness), which introduces internal excitations to the gears that could cause whining noises and possible tooth separations under certain loading conditions.

In Figure 5.13, the dependent variable is the stiffness and the independent variable is the meshing time, from 0 to €tz, where the total contact ratio, tz the meshing period for a single tooth, and åtz the whole meshing period.

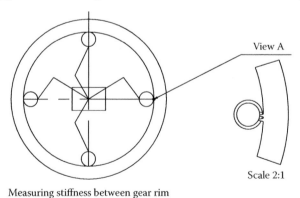

View A

Scale 2:1

Measuring stiffness between gear rim
and pinions planetary reducer.

FIGURE 5.13
Measuring stiffness variation.

5.10.5 Optimizing Overall Durability Performance

Wind turbines are designed to cover a 20-year lifetime with low operating and maintenance costs, and to withstand the high variability of wind forces. To meet these targets, engineering teams apply durability analysis using loads measured through testing or obtained from dynamic simulation. The critical loads acting on a wind turbine are mainly due to fluctuations in speed and direction of the wind and by the starting and stopping of the system. The number of load cases typically varies between 50 and 500 and last up to 10 minutes each.

LMS Virtual.Lab Durability provides the batch capability to efficiently analyze the different load cases independently or on remote CPUs with advanced monitoring capabilities and e-mail notifications at the end of the job. Starting from the external loads, Virtual.Lab computes internal loads acting on the components, such as the rotor hub (Figure 5.14). A global load spectrum formulated from individual events is then used to simulate 20 years of use.

Traditionally, only the static loads and a maximum stress criterion are applied, not taking into account the unpredictable random forces, or the interrelationships of multiaxial forces. This oversimplification leads to inaccurate, fatigue-life predictions, and the only way to secure the operational safety of the turbine was to over-design it. By ignoring the phasing relations, the conclusions may even identify the wrong hotspots.

LMS Virtual.Lab Durability provides accurate fatigue life predictions and results are calculated very efficiently. The time history approach uses all significant damaging events and automatically accounts for correct phase

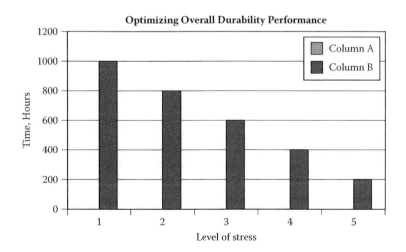

FIGURE 5.14
Optimizing overall durability performance.

relations of load components and correct mean stresses. Because it can assess the entire model of the structure, hotspots are accurately and automatically identified, and the damage distribution is easily visualized.

Furthermore, Virtual.Lab Durability provides the tools to understand the causes of fatigue problems (which events are the most damaging, what is the contribution of a load to a given hotspot) and to refine the design of the components. Using the integrated automation scripting methods, the complex setup of several hundred load events is constructed from a spreadsheet, the fatigue analysis is run in batches, and a report is prepared in a fully automated way.

The unique capabilities of Virtual.Lab Durability include frequency-based fatigue solutions that directly use random descriptions of the loads via power spectral densities and phasing information via cross correlations. The software also performs fast simulations of shaker table tests of sine sweeps and block sines.

5.10.6 Complying with Noise Regulations

To comply with the restrictive regulations, wind turbine development teams have to perform noise and vibration analyses. The noise generated in wind turbines consists of broadband and tonal components. Broadband noise mainly originates from aerodynamic phenomena like the flow of air around the blades, hub, and tower. Tonal noise tends to originate from mechanical components and electrical equipment like the gearbox and the generator. The rotation of these components and the resulting dynamic forces cause local housing surface vibrations, which distribute noise to the surrounding area through radiation. The noise generated by driveline rotating machinery also propagates directly through structural noise paths.

While it is relatively easy to predict performance at the component level, most noise and vibration problems are only discovered at the full system level. From the early concept development stages onwards, LMS Virtual. Lab Noise and Vibration captures all critical process steps to systematically improve the noise and vibration characteristics of a complete assembly. The loads can be imported from measurements, from multibody or acoustic simulations, or from generic loading sources. LMS Virtual.Lab Noise and Vibration offers a wide range of visualization and analysis tools to quickly investigate transfer paths and efficiently assess the noise and vibration contribution of individual system parts.

5.10.7 Optimizing the Overall Wind Turbine System Behavior

The performance of advanced mechanical system designs such as wind turbines relies on an optimal interaction of subsystems of a different nature. Wind turbine engineers have to carefully optimize the coupling between mechanical subsystems like the turbine blades and gearbox on one hand, and the electrical generator and the power grid on the other hand. Electronic controllers, mechanical components, and powered actuator subsystems all relate to different physical domains and a different engineering logic.

This makes it very challenging to assess the multidomain, overall system-level behavior of a complete wind turbine system. Dedicated simulation tools can optimize the individual systems, but cannot take into account the interaction with other systems of a different nature. Testing the full system behavior on a wind turbine prototype is often used as a costly and time-consuming solution.

LMS Imagine.Lab AMESim offers a complete 1D simulation platform to model and analyze multidomain, intelligent systems and to predict their multidisciplinary performance. The AMESim software models critical systems, such as the turbine blades, gearbox, electrical generator, and power grid, as analytical or tabulated models. It studies the performance of the individual subsystems and the coupling between the subsystems in the overall system configuration.

The blades model transforms the wind energy into mechanical energy on a rotating axis. The efficiency of the wind turbine, which depends on a reduced speed parameter, is integrated in the model. This efficiency has a bell shape curve. When the wind speed is low, the turbine is not efficient. If the turbine rotates too fast, its efficiency is not optimal either. The AMESim software offers a wide range of modeling levels for the gearbox, from a simple efficiency ratio model up to a detailed model that simulates the teeth backlash with variable contact stiffness. Dedicated libraries are available to model the electrical components like the generator, the inverter, and the power grid. The design of control rule can be done directly in AMESim, or by using Matlab®/Simulink. In the latter case, the control model can easily be linked back into AMESim. The AMESim software allows the engineering team to interactively assemble the overall system model, and to simulate key system-level performance criteria. As an example, the software can simulate the rotating velocity of the wind turbine and assess the electrical power that is generated, as a function of the input wind velocity.

The developed model of the overall intelligent system allows one to:

- Compare the use of a synchronous machine or induction machine
- Compare the use of a mechanical reducer or machine with high number of pole pairs
- Study different alternative control rules
- Optimize energy management
- Detect potential online failures

5.10.8 Conclusions

The integrated simulation capabilities within LMS Virtual.Lab and LMS Imagine. Lab AMESim offer an efficient solution to analyze and optimize the dynamics, durability performance, and noise emissions of wind turbines. Accurate loads are easily generated with LMS Virtual.Lab Motion thanks to state-of-the-art contact formulations suited for system-level analysis. Those loads lead the engineers to evaluate the stresses occurring in each component and the vibrations generated in the structure. To assess the fatigue life of the components, either the computed stresses or measured stresses are then used in Virtual.Lab Durability. Virtual.Lab Noise and Vibration helps engineers to evaluate the noise emitted by the system and to understand its components and origin.

Using the same LMS Virtual.Lab environment to perform all of these analyses eliminates the need to transfer data and models between different tools, which saves time and avoids errors. Moreover, the single integrated environment enables one to quickly analyze the effect of design changes on a specific performance attribute. This allows engineering teams to perform fast optimization loops from the early development stages onward.

The multidomain system approach of LMS Imagine.Lab AMESim helps engineers to model critical subsystems, to study their performance, and to assess the coupling between the subsystems in the overall system configuration.

References

1. Golfman, Y., L. P. Rochkov, and N. P. Sedorov N. P. 1978. *Polymer materials application on the rotor blades for hovercrafts.* Krylov, Russia: Central Scientific Research Institute.
2. Golfman Y., 1977. Polymers for fiberglass protection. Russian Patent 594745.
3. Leder, B. 2003. World's first civil tiltrotor achieves first flight. Press release. Bell Helicopter, March 7. Online at: www.belhelicopter.com/companyinfo/pressReleases/pr_0307001.html
4. Bonassar, M. J. 1980. *MM&T fiber-reinforced plastic helicopter-tail rotor assembly (pultruded spar).* Final report (August 1975–October 1978). Ft. Rucker, AL: US Army Aviation R&D Command.
5. Smith E. C. 1994. *Vibration and flutter of stiff-inplane elasticity tailored composite rotor blade mathematical and computed modeling.* Rotorcraft Modeling (Special edition) 20 (1, 2).
6. Chou, P. C., and N. J. Pagano. 1992. Elasticity tensor, dyadic, and engineering approaches. New York: Dover Publications.
7. Golfman, Y. 1966. The interlaminar shear stress analysis of composite in marine front. Paper presented at the 32nd International SAMPE Technical Conference, Arizona.
8. Golfman, Y. 2003. Dynamic aspects of the lattice structure behavior in the manufacturing of carbon-epoxy composites. *Journal of Advanced Materials* 35 (2).

9. Golfman Y. 2001. Fiber draw automation control. *Journal of Advanced Materials* 34 (2).
10. Kushul, M. Y. 1964. *The self-induced oscillations of rotors*. New Brunswick, NJ: Consultants Bureau.
11. Golfman, Y. 1993. Ultrasonic non-destructive method to determine modulus of elasticity of turbine blades. *SAMPE Journal* 29 (4).
12. Hou, A. and K. Gramoll. 2000. Fabrication and compressive strength of the composite attachment fitting for launch vehicles. *Journal of Advanced Materials* 32 (1).
13. Gibilisco, S. 2002. *Physics demystified*. New York: McGraw-Hill.
14. Golfman, Y., E. K. Ashkenazi, L. P. Roshov, and N. P. Sedorov. 1974. *Machine components of fiberglass for ship building*. Leningrad, USSR: Shipbuilding Institute.
15. Zlochevskii, A. B., and V. I. Mironov. 1969. Measuring deformations of highly elastic materials with resistance strain gauges. *Measurement Technique* 2: 38–39.
16. Golfman, Y. 1969, *Influence technology pressing fiberglass blades at their strength*. Leningrad, USSR: Shipbuilding Institute.
17. Golfman, Y. 2010. *Hybrid anisotropic materials for structural aviation parts*. Boca Raton, FL: Taylor & Francis Group.
18. EC Project No. EN3W.0024, 1987–1989. Design basics for wind turbines.
19. EC-JOULE-1 programs, 1990–1993. Fatigue properties and design of wind blades for wind turbines. Contract # JOUR-0071-DK (MB).
20. EC-JOULE-1 programs, 1992–1995. Development of advanced blades for integration into wind turbine systems. Contract # JOU2-CT92-085.
21. Kensche, C. W. 1996. Fatigue of materials and components for turbine rotor blades. Luxembourg: EUR 16684, European Community.
22. Mayer, R. M. 1996. *Design of composite structures against fatigue*. Suffolk, U.K.: Mechanical Engineering.
23. Smet, B. J., and P. W. Bach. 1994. Database FACT: Fatigue of composites for wind turbines. Paper presented at the Third IEA Symposium, Wind Turbine Fatigue, ECN, Petten, The Netherlands, April 22–23.
24. Mandell, J. F., and D. D. Samborsky. 1997, *Composite materials fatigue database. Test methods. Materials and analysis*, SAND97-3002, UC-1210. Albuquerque, NM: DOE/MSU, Sandia National Lab.
25. DOE/MSU. 2004. *Composite material fatigue database*. (Update: February 24) Online at htpp://www.sandia.gov/RenewableEnergy/wind.energy/other/973002upd0204
26. EU FP5. 2001–2006. Reliable optimal use of materials for wind turbine rotor blades (OPTIMAT BLADES). ENK6-CT-2002-00552. Online at: http://www.vmctudelft.nl/optimat.blades/
27. van Wingerde, A. M., R. P. L. Nijssen, D. R. V. van Delft, L. G. J. Jansen, P. Brondsted, et al. 2003. *Introduction to the OPTIMAT BLADES project*. in Proc. CD-ROM.CD, www.wmc.eu/public~docs/index.htm
28. European Wind Energy Exhibit. 2003. European Wind Energy Association Conference, Madrid, Spain, June 16–29.
29. Megnis, M. P. Brondstead, S. A. Rehman, and T. Ahmat. 2004. Life prediction of long fiber composites in extreme environmental conditions using damage evolution approach. Paper presented at the Proceeding of the 11th European Conference Composite Materials. Rhodes, Greece. London: European Society of Composite Materials.

30. Smits, A., D. van Hemelrijick, and T. Philippidis. 2004. The digital image correlation technique as full field strain technique on biaxial loaded composites using cruciform specimens. Paper presented at the 12th International Conference of Exp. Mech. Politecnico di Bari, Italy.
31. Smits, A., D. van Hemelrijick, T. Philippidis, A. M. van Wingerde, and A. Cardon. 2004. Optimization of a cruciform test specimen for bi-axial loading of fiber reinforced material systems. Paper presented at the Proceedings of the 11th European Conference of Composite Materials. Rhodes, Greece. London: European Society of Composite Materials.
32. Johnson, D. P. 2001. Thermal cracking in scaled composite laminates. *Journal of Advanced Materials* 33 (1).
33. Bailey, J. E., P.T. Curtis, and A. Parvizi. 1979. On the transverse cracking and longitudinal splitting behavior of glass and carbon fiber reinforced epoxy. Paper presented at the Proceedings of the Royal Society, London, A366, 599–623.
34. Talreja, R. 1987. *Fatigue of composite materials.* Technical University of Denmark: Technomic Publishing Co.
35. Brondsted, P., H. Lilholt, and A. Lystrup, 2005. *Composite material for wind power turbine blades.* Roskilde, Denmark: Annual Review, a nonprofit scientific publisher. email: powlbrondsted@risoe.dk, hans.lilholt@risoe.dk, aage.Lustrup@risoe.dk
36. Golfman, Y. 1991. Strength criteria for anisotropic materials. *Journal of Reinforced Plastics and Composites* 10 (6).
37. Golfman, Y. 2004. The fatigue strength prediction of aerospace components using reinforced fiber/glass or graphite/epoxy. *JAM* 36 (2).
38. Parton, V. Z., and P. I. Perlin. 1984. *Mathematical methods of the theory of elasticity,* Moscow: MIR.
39. Hodgman C. D. 1959. *CRC Standard Mathematical Tables*, The Chemical Rubber Publishing Company, Cleveland, Ohio, 285.
40. LMS Virtual Lab. 2003. The Integrated Environment for Functional Performance Engineering. http://novicos.de/system/files/files/76/virtual-lab.pdf
41. G. Lethe, J.D. Cuper, Simulating dynamics, durability, and noise emission of wind turbines, 2009,v.23, No.4, www.springelink.com
42. Cai, Y. 1995. Simulation on the rotational vibration of helical gears in consideration of the tooth separation phenomenon (a new stiffness function of helical involute tooth pair). *The ASME Journal of Mechanical Design* 117: 460–469.
43. Cai, Y., and T. Hayashi. 1994. The linear approximated equation of vibration of a pair of spur gears (theory and experiment). *The ASME Journal of Mechanical Design* 116: 558–564.
44. International Organization of Standardization. ISO Standard 6336-1, Calculation of load capacity of spur and helical gears, Part 1: Basic principles, introduction and general influence factors. Online at: www. iso.org.

6

NDE Digital Methods for Predicting Stiffness and Strength of Wind Turbine Blades

6.1 Ultrasonic Nondestructive Method to Determine Modulus of Elasticity of Wind Turbine Blades

6.1.1 Introduction

The ability of ultrasonic waves to travel in a web direction over a minimum time for composites was advanced by the author in 1966.[1] This effect can be used for different applications in composites, ceramics, and metal alloys.[2,3] Nondestructive evaluation (NDE) of the material properties of a structure makes this method very useful. Research has been conducted to determine the modulus of elasticity of high-speed turbine blades.[4] In this research, a general nondestructive test method for deterring the modulus of elasticity in different directions was examined.

Modules of elasticity predictions of turbine blades using an ultrasonic method probably give the option of estimating the strength of the blades. Gershberg[5] determined the parameters of elasticity for specimens fabricated from fiberglass. Golfman and Gershberg determined the parameters of elasticity in the blades of screw propellers fabricated from orthotropic fiberglass using a nondestructive ultrasonic method.[6] The advantage of a nondestructive evaluation of the properties of new materials used for manufacture of turbine blades is that they can be predicted with little difficulty.

6.1.2 Theory and Application of Ultrasonic Method

Knowledge of the parameters of elasticity is necessary to calculate stress and estimate the strength of turbine blades. If we assume that composite turbine blades to be manufactured from orthotropic material having three planes of elasticity symmetry, its technical behavior can be completely characterized by nine classic constants.[7]

The elastic constants for orthotropic materials are listed in Table 6.1.

For a composite turbine blade, there are nine independent parameters: E_z, E_y, and E_x are the modulus of elasticity in the three principal directions z,

TABLE 6.1

The elastic constants for orthotropic materials

$1/E_z$	$-\mu_{yz}/E_y$	$-\mu_{xz}/E_x$	0	0	0
$-\mu_{zy}/E_x$	$1/E_y$	$-\mu_{xy}/E_x$			
$-\mu_{zx}/E_z$	$-\mu_{yx}/E_y$	$1/E_x$			
			$1/G_{zy}$		
				$1/G_{yx}$	
					$1/G_{xz}$

y, and x, respectively; G_{zy}, G_{yx}, and G_{xz} are the modulus of shear in the ply orientation zy, and the interlaminar directions are zx and yx. μ_{zy}, μ_{yx}, and μ_{xz} are the Poisson's ratios. The first letter in the subscript of μ represents direction of force applied and the second letter represents the transverse direction of E_{45}, is the Young modulus along $45°$ in the plane zy, and μ_{45} is the Poisson ratio along $45°$ orientation.

$$G_{zy}=E_{45}/2(1+\mu_{45}) \tag{6.1.1}$$

All nine parameters can be predicted using the ultrasonic nondestructive method. This method is based on the measure of the time interval of ultrasonic oscillations in the longitudinal and the transverse directions. Gershberg[5] used semiconductors of model YKC-1 to measure the time interval. Semiconductors YKC-1 are provided with ultrasonic heads having frequencies ranging from 20 to 240 kHz. To maintain acoustic contact between the ultrasonic heads, the blades, and the blades' surface, either viscous liquid or highly viscous oils can be used in the immersion medium. The acoustic heads are installed on the surface of the blade as shown in Figure 6.1. Along the longitudinal direction, a wave of 80 kHz was used to measure the time of interval. A 100 mm circle diameter is marked on the surface of the blade. The position of the acoustic heads is varied with each frequency and time interval of measurement (see Figure 6.1).

The objective is to find the direction in which the ultrasonic wave can travel in a minimum of time. The velocity of the ultrasonic wave can be calculated using the following expression:

$$C=L/t * 10^3 \tag{6.2}$$

where:

C is the velocity of the ultrasonic oscillations;

L is the length between the two acoustic heads;

t is the time taken for the ultrasonic oscillations to reach from one head to the other.

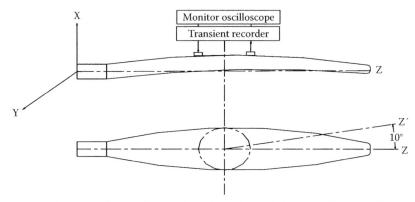

Schematic diagram showing the position of acoustic heads installed on surface of wind turbine blades. Ultrasonic wave test method application.

FIGURE 6.1
Ultrasonic wave test method: Acoustic heads were installed at surface of the wind turbine blade.

The equation to calculate the in-plane modulus of elasticity in any direction for an infinitely long plate is obtained for angle f. We solve the different equations of motion and strain are correlation in an orthotropic body.

$$E\phi = C_\phi \rho^* (1 - \mu_{1\phi}\mu_{2\phi}) \tag{6.3}$$

where:
  ϕ is the angle in the direction of the zy plane;
  C_ϕ is the velocity in the direction of the angle ϕ;
  $\mu_{1\phi}\mu_{2\phi}$ is the Poisson's ratio in the direction of the angle ϕ.

The fiber composite has high strength, stiffness, and low weight that can be tailored to be applicable, and the reinforcing fibers can be oriented in the matrix to provide strength in any direction. From Equation (6.2) and Equation (6.3), the material properties of an infinitely long plate are computed as follows:

$$\mu_{ZY} = \left(\frac{C_z}{C_y}\right)\mu_{YZ} \tag{6.4}$$

$$E_y = \frac{10\rho C_y^{\,2}}{g}\left(1 - \frac{C_z^{\,2}}{C_y^{\,2}} * \mu_{yz^2}\right) \tag{6.5}$$

$$E_z = \frac{C_z^2}{C_y^2} * E_y \qquad\qquad (6.6a)$$

$$E_{45} = \left(-\frac{g}{10\rho C_{45}^2 A^2} - \frac{2}{A} \right) \qquad\qquad (6.6b)$$

$$\mu_{45} = 1 + E_{45}*A \qquad\qquad (6.7)$$

$$A = \frac{\mu_{ZY}}{E_z} - \frac{1}{2E_z} - \frac{1}{2E_y} \qquad\qquad (6.8)$$

$$Gzy = \frac{E_{45}}{2(1 + \mu_{45})} \qquad\qquad (6.9)$$

where:

C_z, C_y, C_{45} are the velocities of longitudinal ultrasonic oscillations of the infinitely long plate in the warp, fill, and diagonal directions (m/sec);

g is the acceleration due to gravity (9.81 m/sec^2);

ρ is the density of the composites (1.998 g/cm^3).

The z direction set is called the *warp* and the y direction set is called the *fill*. To determine elastic properties of a composite material, 12 plates having geometrical dimensions of 10 × 200 × 250 mm were fabricated containing 20 to 30 percent epoxy and 70 to 80 percent glass fiber and under pressure 100 kg/cm^2, and temperature of 160°C, time of curing 3 to 6 min/mm. The fabricated plates, with defined warp and fill and diagonal directions, were cut into 10 × 10 × 200 mm and 10 × 30 × 200 mm specimens.

The elastic properties E_z, E_y, E_x, E_{45}, G_{zy}, G_{xz}, G_{yz} for these specimens were determined using the ultrasonic method.

These properties are transferred from an infinitely long plate to finite dimensions of the turbine blade using appropriate coefficients given below:

$$K_z = \frac{E_z^t}{E_z^u}; \quad Ky = \frac{E_y^t}{E_y^u}; \quad K_x = \frac{E_x^t}{E_x^u} \qquad\qquad (6.10)$$

where the superscript t denotes the standard destructive test and u denotes ultrasonic test. The elastic properties in the turbine blades fabricated from composites are given by:

$$\mu_{ZY} = \frac{K_z}{K_y}\left(\frac{C_z}{C_y}\right)\mu_{ZY} \tag{6.11}$$

$$E_y = \frac{K_y}{K_z}\frac{10\rho C_y^2}{g}\left(1 - \frac{C_z^2}{C_y^2}*\mu_{yz}^2\right) \tag{6.12}$$

$$E_y = \frac{K_z}{K_y}\frac{C_z^2}{C_y^2}*E_y \tag{6.13}$$

$$E_{45} = K_{45}\left(-\frac{G}{10\rho C_{45}^2 A^2} - \frac{2}{A}\right) \tag{6.14}$$

The coefficient values for the turbine blades were found to be $K_z = 0.885$, $K_y = 0.840$, and $K_{45} = 0.8$, where Cz, Cy, and C_{45} are velocity propagations of longitudinal oscillations in turbine blades in the warp, fill, and diagonal directions. Equation (6.7) to Equation (6.9) can be used for calculation of μ_{45}, A, and G_{zy} for a turbine blade. Application of the ultrasonic wave method for turbine blades is demonstrated in Figure 6.1. A circle drawn on the blade surface and ultrasonic heads was aligned with the circumference of the circle along the geometrical axis. The time taken for the wave to pass from one head to another was measured. Then, the heads were moved away from the geometrical axis by 5 degrees and the propagation time was again measured. The results are shown in Table 6.2.

To achieve accurate results, it is necessary to take the mean $\bar{X}$ and the quadratic deviation $\bar{S}$ (Table 6.3).

From an analysis of the data for the 42 specimens, the values for $X = 3.56*10^5$ and $S = 0.25*10^5$ were obtained. Figure 6.2 to Figure 6.4 show the experimental and theoretical distribution modulus curves of elasticity Ez, Ey, and Ex.

TABLE 6.2

Velocity Propagation of Longitudinal Waves in the Blades

Blade No.	Base, mm	Time, sec	Velocity C_z, m/sec	Time, sec	Velocity, C_y m/sec	Time, sec	Velocity, C_{45}, m/sec
1	100	23.0	4350	25.7	3900	26.2	3800
2	100	23.0	4350	25.0	3950	26.0	3850
3	100	23.5	4250	26.0	3850	26.6	3750
4	100	23.0	4350	25.7	3900	26.2	3800
5	100	22.4	4450	25.7	3900	26.2	3800

TABLE 6.3

Results of Static Treatment to Determine Modulus of Normal Elasticity E_z

E_z*103 kg/ cm²	Empirical Frequency	'x *105	'Sx*105	'x- 'x1 s	'(x) eq. 15	m_1 m_1	c^2. eq. 16
3.1 3.15 3.2	5	3.56	0.25	1.63	0.0035	3—2	0.133
3.25 3.3 3.35	7			1.04	0.139	6—1	0.166
3.4 3.45 3.5	8			3.56	0.217	9— -1	0.101
3.55 3.6 3.65	10			0.016	0.240	10—0	0
3.7 3.75 3.8	8			0.76	0.18	8—0	0
3.85 3.9 4.0	4			1.35	0.096	4—0	0

Note: N = å42; c2 = å0.4

We assumed that the empirical curve follows the law of normal distribution.[8] The normal law of distribution is given by:

$$\rho(x) = \frac{1}{(2\pi 6)^{1/2}} \, \bar{e} \, \frac{(\bar{x}-a)^2}{2\sigma^2} \tag{6.15}$$

where $\bar{X}_i$ and S are substituted instead of "a" and "6," respectively. With Equation (6.15) and by substituting experimental results, we can check whether they follow the law of normal distribution. We assumed that the empirical curve follows the theory of normal distribution if the reliability testing is not less than 95 percent. Therefore, the probability is more than 5 percent (0.05). To compare the experimental and theoretical curves, density of probability by Pearson's criterion is used, which is given by the *Methods of Statistical Treatment of the Experimental Datum.*[9]

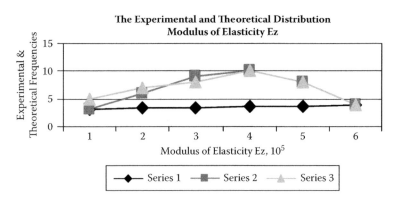

FIGURE 6.2
Modulus of elasticity Ez distribution.

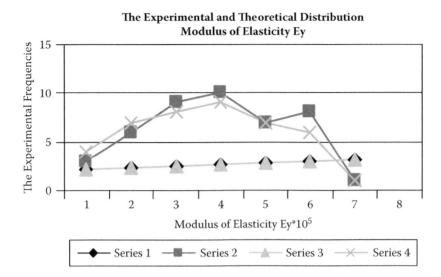

FIGURE 6.3
Modulus of elasticity Ey distribution.

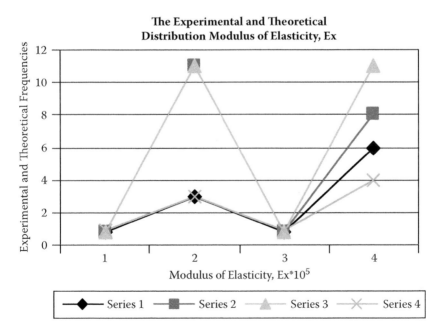

FIGURE 6.4
Modulus of shear elasticity Gzy distribution.

$$\chi^2 = \sum_{I=1}^{n} \left(\frac{m_i m_i{'}}{m_i} \right)^2$$ (6.16)

where:
   m_i is the experimental frequencies;
   $m_i{'}$ is the theoretical frequencies.

After determining χ^2 from Equation (6.16), we determine coefficient K using the formula k = n-r-1.
Here:
   k is the number of degree of freedom;
   n is the number of comparison frequencies;
   r is the number of parameters in the theoretical function of distribution.

The normal law of distribution has two parameters. In our case, we assume worth variant when empirical frequencies are equal to 7, $X^2 = .166$ (see Table 6.3). Following the last equation, when r = 2, n = 7, and k = 4 from the table of probabilities,[9] we find $P(x^2) = 0.9098$. The experimental data appears to agree with the theoretical curve. Similarly, the experimental values of Ey, and Gzy specimens were obtained and checked whether they satisfy the Pearson's criterion. For E_y, the values obtained were $\overline{X} = 2.595* 10^5$ and $\overline{S} = 0.232*10^5$, and for G_{zy} the values were $\overline{X} = 0.818*10^5$ and $\overline{S} = 0.0172*10^5$. The Pearson's criterion for both E_y and G_{zy} is obtained as 0.9098, which implies that the experimental values agree by over 90 percent with the theoretical values. The values of modulus of elasticity and modulus of shear for turbine blades are listed in Table 6.4.

The modulus of elasticity in angle direction can be found as:

$$E\phi = \frac{E_z \lambda}{\lambda \cos^4 \phi + 2B \cos^2 \phi + \sin^4 \phi}$$ (6.17)

where:

$$\lambda = \frac{E_y}{E_z}; \quad 2B = 4 \frac{E_y}{E_{45}} - (1 + \lambda)$$

TABLE 6.4

Significances of Normal Modulus and Shear Elasticity in the Turbine Blades

Parameters	E_z*10^5 kg/cm^2	E_y*10^5 kg/cm^2	$E_{45}*10^5$ kg/cm^2	$G_{zy}*10^5$ kg/cm^2	$G_{zx,yx}$	μ_{yz}	μ_{zy}	μ_{45}
Specimens	3.55	2.59	2.2	0.18	0.68	0.1	0.13	0.34
Blades	3.4	2.56	2.28	0.18	0.68	0.1	0.13	0.34

The velocities of propagation of longitudinal waves in different angles with a step of 15° are determined using Equation (6.17). From these velocities, the modulus of elasticity is computed using Equation (7-9) and Equation (11-14). The values of the coefficient introduced into Equation (2) are:

$$E\phi = k\phi C\phi^2 \rho(1 - \mu_{1\phi}\mu_{2\phi}) \tag{6.18}$$

where the $k\phi$ are computed by:

$$k\phi = \frac{k_z}{\cos^4\phi + b_o\cos^2 2\phi + c_o\sin^4\phi} \tag{6.19}$$

where:

$$c_o = \frac{k_z}{k_y}; \quad bo = \frac{k_z}{k_{45}} - \frac{c_o + 1}{4}$$

The difference between the values computed by Equation (6.17) and Equation (6.18) is only about 5 percent, which proves that the values obtained by the present nondestructive method are acceptable (see Table 6.5).

Intent coefficients of thermal expansion were in the radial direction $\alpha_1 = .118$ cm/cm/°C /$3.8*10^{-3}$ in/in/F; in tangential direction $\alpha_2 =. 115$ cm/cm°C/$3.71*10^{-3}$ in/in/F; intent coefficient of thermal expansions in an arbitrary direction were determined as $\alpha_{12} = 2\sin\theta*\cos\theta(\alpha_1-\alpha_2)$.

6.1.3 Conclusions

1. Turbine wind blades applications utilize hybrid reinforced composites for the huge blades that replace traditional fiberglass blades.

TABLE 6.5

Modulus of Normal Elasticity Depends on Angles

Angle of Degree	μ_ϕ	k_ϕ	$C\phi$ m/sec	$E\phi*10^5$, kg/cm² Eq. (6.17)	$E\phi*10^5$, kg/cm² Eq. (6.18)	Error,%
0	0.13	0.855	4350	3.4	3.3	3.0
15	0.20	0.86	4150	3.1	2.9	5.0
30	0.30	0.835	3950	2.6	2.5	4.0
45	0.34	0.80	3800	2.28	2.2	3.5
60	0.13	0.812	3900	2.29	2.38	4.0
75	0.128	0.824	3940	2.48	2.4	3.5
90	0.10	0.840	3900	2.56	2.46	4.0

2. Nondestructive digital test methods are necessary because all destructive testing is very expensive.

3. In this research, we developed a general nondestructive digital test method for determining the modulus of elasticity of the wind turbine blades. Reliability results of more than 90 percent prove this method, which is very useful for green alternative technologies.

6.2 Dynamic Local Mechanical and Thermal Strength Prediction Using NDT for Material Parameters Evaluation of Wind Turbine Blades

6.2.1 Introduction

As stated above, the ability of ultrasound waves to travel in a web direction over a minimum of time for curing fiberglass was advanced by the author in 1966.[1] This effect can be used for different applications, such as composites, ceramics, and metal alloys. However, only the combination of thermographs, ultrasound, and radiography techniques can predict the physical parameters of a material (density, thickness, modulus of elasticity, and strength).

The properties of carbon—carbon composites are shown in Table 6.6. The parameters of the relationship to strength for carbon—carbon composites are shown in Table 6.7.

The overall approach is to determine the local elastic constant and material parameters using nondestructive methods, which characterize strength in the domain point and also determine the local stresses from the temperature profile.

The purpose of this research is to show and discuss the developed NDE methods used to increase the reliability estimation of dynamic strength.

The thermal, radial, and tangential stresses for an orthotropic nose were used as an example while the dynamic modulus of elasticity for stress analysis was used and all the stress components for strength criteria were established.[4,10,11]

Agfa Nondestructive Testing, Inc. (Lewistown, Pennsylvania), after successfully testing caps for three years using the ultrasound method, has become a leading supplier of NDT systems.[12]

Composites structures for space programs have been mainly manufactured from sandwich composites with aluminum honeycomb and graphite/epoxy face sheets and are located on the upper part of the space vehicle launcher where their influence has the largest effect.[13]

For example, the influence rates of composites on different stages varies on the Ariane 5 rocket from 7 percent for the lateral booster to 100 percent for the upper parts. They require that a strong effort be made in the design of structures in the launcher upper part and the demand for NDT equipment and procedures becomes imperative.

TABLE 6.6

Properties for Carbon–Carbon Composites

Modulus, of elasticity acting in x,z directions MPa/psi	Normal strength, of thermal expansion MPa/ psi	Shear strength MPa/psi	Coefficient cm/cm Co				
Significance	E11	E22	s_{b1}	sb2	t_{b12}	a_1	a_2
Compression	.8*105/ 12,3*106	.65*105/ 9.5*106	135.8/ 19.700	116.5/ 16.900	9.37/ 1360	.118	.115
Tension	.83*105 12.1*106	.71*105/ 10.3*106	226/ 32.800	172/ 24.970	9.37/ 1360	.118	.115

TABLE 6.7

Parameters Relationship to Strength for Carbon–Carbon Composites

Significance	c	b	d	s	h
Compression	1.16	12.43	14.48	-12.34	-10.87
Tension	1.31	18.36	24.1	-21.86	-17.17

Launched in April 2001, NASA's Mars Odyssey is now prepared to collect data that will offer insights into the makeup and history of the Red Planet. With the spacecraft's 20-ft boom successfully deployed, two neutron detectors and the gamma ray spectrometer (GPS) mounted on its end, the spacecraft can measure the quantity and distribution of primary elements located at or near the planet's surface and also the modulus of elasticity of the primary elements.

Additionally, silicon, oxygen, iron, magnesium, potassium, aluminum, calcium, sulfur, and carbon are among the 20 primary elements being measured.

NASA's Marshall Space Flight Center (MSFC) has tested Raytheon's Radiance infrared camera to devise NDE methods for assembling assorted aerospace components.

These NDE techniques had been used for any structural anomaly determinations for composites on the Space Shuttle nose cap and the Radiance system was used for the thermography component because the camera's 256 × 256-pixel InSb (indium antimonide) standing focal plane array generates high-resolution images and is highly sensitive to slight temperature changes.[14]

Thermoforming technology includes prepreg layup layers, a curing process with monitoring parameters, pressure, temperature of polymerization time, and a cooling process with a temperature gradient of more than 10°F (12°C). This means that there is an irregular field of temperature and thermal stresses and can be detected by NDT methods.

Newly developed braiding technology for aerospace components has used carbon–carbon or graphite epoxy dry fabric, injected epoxy, and curing was shown by thermal cameras. Cooling processes were not free of thermal stresses and were also seen by NDT methods.

Composite structures had a low thermal conductivity during the faster heating process and a high gradient field of temperature changes. Additionally, an irregular field of temperature had significant thermal stresses, which could result in failure in structures in the process of fabrication. These also were seen by NDT.

Curing and cooling processes for high thickness structures with a low speed of heat up rate can avoid a significant thermal stress, but they increase time and labor costs.

The task of determining the optimal regime is sophisticated because, in the process of curing epoxy matrix, the exothermic reaction appears in adiabatic conditions, which avoid contact with the outside environment. All the heat is used in preheating the epoxy resin.

Finally, the velocity of reaction increases by the exponent of temperature and self-heat in these conditions and could delaminate graphite epoxy composites.

Theoretical investigation of this subject had been done by Golfman.[15]

Dynamic stresses for orthotropic components could be described as:

$$\sigma x = \frac{d^2\phi}{dx^2}; \quad \sigma y = \frac{d^2\phi}{dy^2}; \quad \tau xy = \frac{d^2\phi}{dxdy}; \qquad (6.20)$$

Here: x,y are the vectors describing the directions in which the dynamic stresses act; ϕ, the stress function, can be shown as:

$$\phi = Qij\ \varphi \qquad (6.21)$$

Qij are the nine stiffness constants described in Equation (6.21). φ is the contour of profile for the orthotropic parts.

Thus, Equation (6.21) is

$$\sigma x = \frac{d^2\phi}{dx^2} = h\rho C^2{}_x$$

$$\sigma y = \frac{d^2\phi}{dy^2} = h\rho C^2{}_y \qquad (6.22)$$

$$\tau xy = \frac{d^2\phi}{dxdy} = h\rho C^2{}_{xy}$$

where:
  h is a parameter of length propagation of the ultrasonic wave;
  ρ is the density of composites (for fiberglass 1.998 g/cm³);
  C is the velocity of ultrasonic propagation(m/sec).

By replacing the stress function from Equation (6.21), the results are seen in Equation (6.23):

$$\sigma x = Q_{11} \frac{d^2\varphi}{dx^2} = Q_{11}h\rho C^2_x$$

$$\sigma y = Q_{22} \frac{d^2\varphi}{dy^2} = Q_{22}h\rho C^2_y \qquad (6.23)$$

$$\tau xy = Q_{12} \frac{d^2\varphi}{dy^2} = Q_{12}h\rho C^2_{xy}$$

If the mechanical deformation is equal to zero, it is only considered under thermal stresses.[11]
The equation for thermal stresses would become:

$$\sigma_{ij} = Q_{ij}\, \alpha_{ij}\, T \qquad (6.24)$$

where:
  α is the coefficient of temperature expansion;
  T is the temperature gradient.

Q_{ij} are the stiffness constants that can be determined using an NDE method for velocity propagation.[4]
For the orthotropic nose, thermal radial stresses are:

$$\sigma_{11} = -Q_{11}\alpha_{11}T = \frac{E_1}{1 - \mu_{12}\mu_{21}}\alpha_{11}T$$

while the thermal tangential stresses are:

$$\sigma_{22} = -Q_{22}\alpha_{22}T = \frac{E_2}{1 - \mu_{12}\mu_{21}}\alpha_{22}T \qquad (6.25)$$

Here:
  E_1, E_2 are the moduli of normal elasticity in the radial and tangential directions;

μ_{12} μ_{21} are Poisson's ratio of material—the first symbol designates the direction of force, the second symbol designates the direction of transverse deformation;

α_1 α_{22} are the coefficients of thermal expansion in the radial and tangential directions;

T is the temperature gradient.

The differential equation of heat conductivity without the exothermic reaction of curing of the nose cap is as follows:

$$\frac{dT}{dt} = \beta\left(\frac{d^2T}{dr^2} + \frac{1}{r} * \frac{dT}{dr}\right)$$

(6.26)

where:

t is the time of curing;

R, r are the outside and middle radius of the nose cap;

β is the coefficient of thermal conductivity.

Thus, in selecting the boundary conditions for Equation (6.26), $T(r,0) = 0$; $T(R,t) = bt$; b is the velocity of curing (cooling) process.

The first approach to the solution of Equation (6.27) is:

$$T(r,t) = \frac{bR^2}{\beta} * \left[\frac{\beta t}{R^2} - \frac{1}{4}\left(1 - \frac{r^2}{R^2}\right)\right]$$

(6.27)

The gradient of temperature T during the period of curing (cooling) may be responsible for the geometrical parameters of the nose cap, the thermal conductivity of the epoxy resins, and the velocity of curing b.

We also must look at the stress components for a rotating shell that can be designated following Timoshenko's[16] studies and results.

Here:

R_o, R_i, r are the outside, inside, and middle radius of the nose cap;

ρ is the density of material;

ω is a velocity of rotation and replaced in Equation (6.27) to Equation (6.30).

In Figure 6.1 are wind turbine blades with the installed surface ultrasonic transducers. We used Panametrics Technology contact transducers with low frequencies of 50 KHz (X1021), 100 KHz (X1020) and 180 KHz (X1019). The high voltage pulse receivers, such as Panametrics 5058PR, also were used.

In measuring the temperature gradients, infrared thermometers were used and the intensity of the radiation created by the infrared camera was a function of the temperature gradient. The infrared thermometers simply measured the intensity of radiation and, thereby, measured the temperature. The infrared camera had a 0.025°C sensitivity and was able to detect anomalies, such as delaminations. The influence of the velocity of curing on the rise of the temperature gradient for three types of epoxy resins is very significant.

The strength of composites can be predicted using second order polynomials (Tsai and Wu,[17] Wu,[18] Hoffman,[19] Hill[20]). Strength criteria of second order is not capable of handling air stream load, particularly for strong anisotropic materials like carbon/carbon or graphite epoxy.

In our work,[21] we developed a strength theory and found the strength criteria in tensor form. This criteria can be used separately for tensile and compressive loads.

The probability of local cracking can be predicted using mathematical models, which include the first and second approach in the law of normal distribution.

$$P(t) = \frac{1 - t^{2}/_{2}}{(2\pi)^{1/2}} \tag{6.28}$$

Here:

parameter $t = \dfrac{\sigma_{bi} - \sigma_{bm}}{S_j}$

σ_{bi} is a current strength in x,y,z direction;
σ_{bm} is a middle strength in x, y, z direction;
S_j is a sample of the standard deviation for each environment via:

$$S_j^2 = \frac{1}{n_j} \sum_1^{n_j} \left(\sigma_{bi} - \sigma_{bm} \right)^2$$

Here: n_j is a number of test samples.

We calculated the middle strength σ_{bm} as:

$$\sigma_{bm} = \frac{1}{n_j} \sum_1^{n_j} \sigma_{bi}$$

Thus, for a single test condition (such as compression strength in the fiber direction), the data were collected for each environment being tested. The number of observations in each environmental condition was n_j where j subscript represented the total number of environments being pooled. If the assumption of normality is significantly violated, the other statistical models should be investigated to fit the data.

Finally, the dynamic strength of construction could be predicted as:

$$\sigma_d = 1 - P(t) \tag{6.29}$$

Matrix of material parameters was described as a fourth-rank polynomial equation. The matrix of strength properties was described in the same manner as a fourth-rank polynomial equation.[21] The load of dynamic response doesn't follow Hook's law and has a nonlinear character.

6.2.2 Experimental Investigation Results

The thermal stresses in the nose cap manufacturing of graphite epoxy or carbon–carbon composites can reach a significant value; however, it can never reach the threshold of failure.

For the experimental data, we selected a nose cap with geometrical dimensions Ro = 10" (.254 m); Ri = 9" (.228 m); r = 9.5" (.241 m). The nose cap was fabricated from graphite epoxy and the material had a density of $\rho = .420*10^3$ kg/m^3. The Poisson ratio was $\mu_{12} = \mu_{21} = .036$.

When the modulus of elasticity was calculated E_1, E_2 for the nose cap, the velocity of rotation changed from 3,627 rad/sec to 4,800 rad/sec and to 7,000 rad/sec. The modulus of elasticity in the radial direction was $E_1 = 3.4*10^{10}$ N/m^2, while the modulus of elasticity in the tangential direction was $E_2 = 2.5*10^{10}$ N/m^2 (see Table 6.8). The coefficient of thermal expansion in the radial direction was $\alpha_{11} = -5.45*10^{-6}$ m/m/°C, while in the tangential direction, it was $\alpha_{22} = -5.34*10^{-6}$ m/m/°C.

A second nose cap was fabricated from carbon–carbon and had a density of $\rho = .548*10^{-3}$ kg/m^3; the modulus of elasticity in the radial direction was $E_1 = 1.74*10^{10}$ N/m^2 and the modulus of elasticity in the tangential direction was $E_2 = 1.45*10^{10}$ N/m^2.

The Poisson ratio was: $\mu_{12} = \mu_{21} = .036$; while the coefficient of thermal conductivity for carbon–carbon β was $-.903*10^{-4}$ m/hrm^2 / °C.

Table 6.8 shows the properties for graphite epoxy composites that were used to calculate the thermal, radial, and tangential stresses in Equation (6.25).

TABLE 6.8
Properties of Elasticity for Graphite Epoxy Composite

Description	Values of Characteristics in N/m2							
	E_{11}	E_{22}	E_{45}	G_{12}	G_{21}	μ_{12}	μ_{21}	μ_{45}
Properties of elasticity on the patterns	3.56*1010	2.59*1010	2.24*1010	.818*1010	.68*1010	.13	.10	.34
Properties of elasticity on the nose cap	3.4*1010	2.5*1010	2.2*1010	.818*1010	.68*1010	.13	.10	.34

TABLE 6.9

Significance of the Thermal Stresses on Nose Cap
Manufacturing From Graphite Epoxy Composites

Temperature, C	Stresses Acting in the Radial and Tangential Directions, N/m²	
	$\sigma_{11}*10^{-5}$	$\sigma_{22}*10^{-5}$
20	2237.0	1611.6
17	2072.7	1493.0
14	1908.2	1374.6
12	1743.7	1256.0
9	1579.2	1137.6

The significance of the thermal stresses on nose cap manufacturing from graphite epoxy composites is listed (Table 6.9).

For carbon–carbon composite, the coefficient of thermal expansions that were in the radial direction is $\alpha_{11} = -5.45*10^{-6}$ m/m/°C and in the tangential direction $\alpha_{22} = -5.34*10^{-6}$ m/m/°C.

The coefficient of thermal expansion in an arbitrary direction was determined as $\alpha_{12} = 2\sin\alpha*\cos\alpha(\alpha_{11}-\alpha_{22})$.

The significance of the thermal stress components manufactured from carbon–carbon for the nose cap changed when the gradient of temperature changed from –14°C to –17°C for the compression zone. Also, the same is true for the tension zone.

Future developments of NDE methods consist of the elimination of thermal stresses and to compare with strength parameters (see Table 6.10).

The failure criteria are also needed for design and for guiding materials improvement.[22]

The surface of the equally dangerous biaxial stress conditions for graphite epoxy composite is shown in Figure 6.5.

All the experimental points are found inside this surface, and for the nose cap there are no dangerous conditions. However, if thermal stresses reach a threshold of failure, the nose cap can collapse, and this means that there

TABLE 6.10

Parameters of Strength For Graphite Epoxy Composites, Nm²

Description	Normal Strength, $*10^{-5}$	Inplane Shear Strength $*10^{-5}$	Interlaminar Shear Strength $*10^{-5}$	Normal Strength Acting in Diagonal Directions, 10^{-5}
Significance	$\sigma_{11}, \sigma_{22}, \sigma_{33}$	τ_{12}	τ_{13}	$\sigma_{12}^{45}, \sigma_{13}^{45}, \sigma_{23}^{45}$
Compression zone	3448,827, 3448	34.4	68.9	2068, 2068, 3448
Tension zone	6896, 1241, 34.4	34.4	68.9	4068, 827, 3620

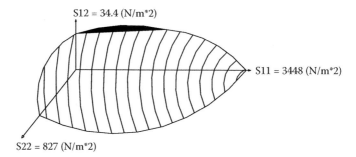

Sigma 11-normal strength in x direction
Sigma 22-normal strength in y direction
Sigma 12-shear strength in 12 directions.

S12 = 34.4 (N/m*2)

S11 = 3448 (N/m*2)

S22 = 827 (N/m*2)

The surface of the equally dangerous biaxial stress
conditions for graphite/epoxy composites.

FIGURE 6.5
Biaxial stresses distributes for graphite/epoxy composites.

TABLE 6.11

Parameters Relationship To Strength For Graphite Epoxy Composites

Significance	c	a	d	e
Compression zone	4.17	.40	100	103.56
Tension zone	5.56	1.7	200	−198.8

are points outside of the surface of strength. It is recommended that the surface of strength be drawn for all the biaxial and triaxial stress conditions. The complex strength values can be calculated if we know all the thermal stresses and all the strength coefficients c, a, d, e (see Table 6.11).

All the calculations in this program were computerized to create a program in C language.

6.2.3 Concluding Remarks

1. A methodology for predicting dynamic strength using the non-destructive evaluation of aerospace components by ultrasound, x-ray, digital radiography, and thermography technologies was established.

2. Thermal stresses for orthotropic composite structures were calculated using parameters, which were found by ultrasound and thermography technologies.

3. Temperature gradients were calculated using the approximate solution of Equation (6.27) fluctuating from −8°C to −16°C. NDE thermography cameras that can observe and record temperature gradients

can be very useful in the process of curing and cooling composite structures.

4. Probability of local cracking can be predicted using statistical models, the first and second approach in the law of normal distribution.

5. In spite of the visibility of cracks, the samples tested reflect 90 percent of initial strength. However, it is very important to use nondestructive evaluation methods in the determination of cracks, which are not visible.

6.3 Noncontact Measurement of Delaminating Cracks Predicts the Failure in Hybrid Wind Turbine Blades

6.3.1 Introduction

Noncontact measurement of delaminating cracks in hybrid wind turbine blades is of great importance in predicting fatigue failure of these blades. Currently used failure prediction methods lack multifunctional self-diagnostic capabilities. In previous studies, the correlation between crack delaminating of the fiber reinforced polymer (FRP) matrix resin and the strain on it was established. However, laminate stress–strain curves are highly nonlinear, especially at elevated temperatures.

In this chapter, we examine the correlation during fatigue testing between temperature gradients and the appearance of nonlinear deformation cracks on the test object.

We also show that measuring the boundary surface temperature gradients of a turbine blade test package can directly self-diagnose and predict the appearance of delaminating cracks, and that injecting a resin agent and solid chemical catalyst can create an automatic self-healing process that increases the durability of the hybrid turbine blades.

6.3.2 Damage Mechanisms of Failure

Recent work by Gamstedt and Talrega[23] describes the fatigue damage mechanisms of carbon FRPs. During fatigue tests, FRPs form delaminating cracks.[24] These authors found a correlation between crack length in a delaminated matrix resin and the applied strain.

We are looking for a correlation between crack length and applied strain in the composites like:

1. Carbon fiber IM7 in polyetheretherketone (PEEK) polyamide
2. Carbon fiber IM7 in polyphenylene sulfide (PPS) polyamide

Layup laminate stress–strain curves are highly nonlinear, especially at elevated temperatures.

We predicted FRP fatigue stress (σ_s) from noncontact measurement of delaminating cracks using Equation (6.30)

$$\sigma_s = \int_1^n E_{11}\Delta\varepsilon^2 e^n \partial n + \int_1 E_{11}\alpha T e^n \partial n \tag{6.30}$$

where:

σ_s is the fatigue stress;

E_{11} is the modulus of elasticity in the fiber-reinforced direction;

$\Delta\varepsilon^2$ is the strain applied *on* the delaminating layer;

n is the number of stress cycles per minute, (change from 1 to 1,000 cycles);

T is the temperature gradient;

α is the coefficient of thermal expansion;

e^n is a exponential function of natural logarithm.

Previous works described how to measure strain ($\Delta\varepsilon^2$) in a resin matrix using embedded fiber optic sensors,[24] and how to measure temperature on the boundary surface of an FRP object[14] with infrared thermography cameras. Our previous work[4] described how to measure the modulus of elasticity using an NDE ultrasonic method. In this chapter, we replace the integrals of Equation (6.30) with summations, Equation (6.31):

$$\sigma_s = \sum_1^n E_{11}\Delta\varepsilon^2 e^n + \sum_1^n E_{11}\alpha T e^n \tag{6.31}$$

Thermography is the use of an infrared imaging and measurement camera to "see" and "measure" thermal energy emitted from an FRP object. Thermal, or infrared energy, is light that is not visible because its wavelength is too long to be detected by the human eye; it's the part of the electromagnetic spectrum that we perceive as heat. Unlike visible light, in the infrared world, everything with a temperature above absolute zero emits heat. The higher Infrared thermography cameras produce images of invisible infrared or "heat" radiation and provide precise noncontact temperature measurement capabilities.

The internal heat of a tested FRP object increases during fatigue testing.[25] If the test frequency is above 1,000 cpm (cycles per minute), the internal heat of the sample typically rises to 50 to 70°C. For test frequencies below 300 cpm, the internal heat decreases to 25 to 30°C. The effect of thermal cycling-induced microcracking in fiber-reinforced polymer matrix composites is studied on the console loading model cantilever beam (Figure 6.6).

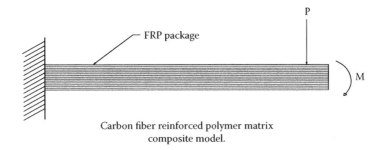

Carbon fiber reinforced polymer matrix
composite model.

FIGURE 6.6
Carbon fiber reinforced polymer matrix composite model. P is an impulse load, M is a bending
moment.

On the level of 1,000 cpm, fatigue stress must be less than fatigue strength
$(\sigma_s < \sigma_{-1})$.

6.3.3   Temperature Measurement of the Surface of an FRP

By defining the ratio σ_s/E_{11} as coefficient K_1, Equation (6.32) becomes:

$$K_1 = e^n(\Sigma\Delta\varepsilon^2 - \Sigma\alpha\, T) \tag{6.32}$$

Table 6.12 and Table 6.13, based on test data from Phoenixx TPC,[26] give K_1
values for two orthogonal fiber orientations of two FRPs:
Table 6.12: K_1 values of IM7/PEEK (Thermo-Lite™) prepreg, made from
carbon fiber IM7 and PEEK polyamide (40 percent and 60 percent by volume, respectively).
Table 6.13: K_1 values of IM7/PPS (Thermo-Lite™) prepreg, made from carbon fiber IM7 and polyphenylene sulfide (PPS) polyamide (40 percent and 60
percent by volume, respectively).
In Equation (6.30), when n is 1,000 cpm, e^n is greater than 2^{1000}, making
K_1/e^n extremely small. If we set this ratio to zero, we obtain this correlation
between temperature gradient in Equation (6.30) and strain.

TABLE 6.12

Mechanical Properties of IM7 PEEK (Thermo-Lite™) Prepreg

Laminate Property	Fiber Orientation	Test Data	K_{-1} Ratio
Fatigue Strength, s_{-1} Ksi	0^0	285	0.0117
	90^0	11.2	0.086
Tensile Modulus, E_{-11}, Msi	0^0	24.2	
	90^0	1.3	
Compression Strength-Ksi	0^0	136	0.0093

TABLE 6.13

Mechanical Properties of IM7/PPS (Thermo-Lite™) Prepreg

Laminate Property	Fiber Orientation	Test Data	K_{-1} Ratio
Fatigue Strength, Ksi	0^0	285	0.0117
	90^0	12.5	0.083
Tensile Modulus, Msi	0^0	24,2	
	90^0	1.5	
Compression Strength, Ksi	0^0	185	0.0108

$$T = \frac{\Sigma \Delta \varepsilon^2}{\Sigma \alpha} \tag{6.33}$$

Table 6.14 gives the modulus of elasticity and coefficient of thermal expansion for the prepreg laminates based on IM7 carbon fibers shown in Table 6.12 and Table 6.13.[26]

The 10 percent of Boron fibers volume fraction added to IM7 fiber. Strain in different layers of the FRP object correlates with each layer.

We assume that correlation between strain $\Delta \varepsilon^2$ during delaminating and number of cracks are

$$\Delta \varepsilon^2 = (n^* C)^{1/2} \tag{6.34}$$

Here:
   n is a number of cracks;
   C is a length of cracks.

Value of parameters are given in Table 6.15.

Figure 6.7 shows the measured correlation between temperature gradient and strain propagation.

Reference 28 shows the test stand used to measure delaminating cracks. Fiber optic sensors are embedded in the FRP package and they detect laser light passing through the material. The infrared camera measures temperature at the side surface boundary of the package. P is impulse load, M_x is a bending moment.

TABLE 6.14

Modulus of Elasticity and Coefficient of Thermal Expansion for the Prepreg Laminates

Material	IM7/PEEK	IM7/PPS
Modulus of Elasticity, Msi	24.2	24.2
Coefficient of thermal Expansion, PPMF	−0.02	−0.06

TABLE 6.15

Value of Technological Parameters

Length delaminating, mm, C	Cracks Number, n	Parameter, C*n	Strain, Δε², %
6	1	6	2.4
12	2	24	4.9
18	3	54	7.3
24	4	96	9.8
30	5	150	12.4

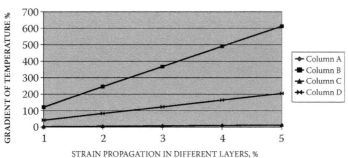

CORRELATION BETWEEN TEMPERATURE GRADIENT AND STRAIN PROPAGATION

FIGURE 6.7
Measured correlation between temperature gradient and strain propagation.

6.3.4 Fatigue Strength Improvement

Figure 6.8 shows the simultaneous injection of a healing agent and catalyst into the test package during stress testing. Injection occurred when a temperature gauge indicated to the injection pumps when the surface temperature of the package increased. The healing agent and solid catalyst are dispersed first.[27] Valves control the four pumps indicate in Figure 6.8.

6.3.5 Conclusions

1. Laser light, detected by imbedded sensors, is a good measure of the formation of delaminating cracks during fatigue testing.
2. An infrared radiation camera accurately measures the surface temperature of test objects during fatigue testing.
3. Surface temperature is a good predictor of fatigue strength as measured by fatigue testing.
4. Injecting a healing agent and a solid catalyst can increase fatigue strength by up to 30 percent.

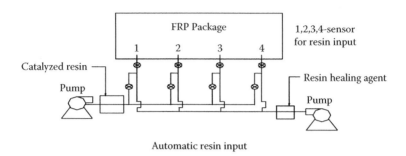

FIGURE 6.8
Automatic resin input into the test package.

6.4 Nondestructive Inspection Technologies for Wind Turbine Blades

6.4.1 Introduction

The turbine wind blades protected with covered luminescent paints will provide visibility under night conditions. All the electronic signals could be transferred from carbon/carbon or carbon/fiber/epoxy composites covering with luminescence paints throughout the components manufacturing without delay.

The protective luminescent paints are currently used to make point measurements on a configuration (propulsion model) for performance or acoustic analysis.

Wind turbine blades described the luminescent paints or thermoplastics layers for nonintrusive measurements of steady and unsteady pressures and temperatures for complex configurations for performance and vibration characteristics.

Pressure sensitive paint (PSP), temperature sensitive paint (TSP), and thin-film MEMS (microelectromechanical systems) are examples of nonintrusive pressure/temperature measurement technologies used for wind turbine design and testing. Such capabilities are desired for underwater applications as well. Capacity sensors measure pressure in propeller blades, which were developed in the compression moulding process.[29]

Propeller blades also have been manufactured from fiberglass, the capacity sensors were used in the pressure range from 50 to 180 kg/cm². We used the shady ultrasound method for the determination defects in the propeller blades (porosity, delaminations, air insertions).

Innovative-Scientific Solutions, Inc. (Dayton, Ohio) developed both the ultrasonic- and pressure-based aerosol deposition techniques for the formation of thin uniform sol-gel-derived PSPs (pressure sensitive paints).[30] The air is a good conductor for transmission of the elastic oscillations. In the air every 10 m, the elevation creates atmospheric pressure of 1, so if the high elevation reaches 1,000 ft., then the acting atmospheric pressure is 100. A new measurement technique for the regime of transonic compressors with PSP was applied in a transonic compressor stage.[31]

6.4.2 Measurement Concept

Thermal, or infrared energy, is light that is not visible because its wavelength is too long (800–1,200 nm) to be detected by the human eye; it's the part of the electromagnetic spectrum that we perceive as heat. Infrared thermography cameras produce images of invisible infrared or "heat" radiation and provide precise noncontact temperature measurement capabilities.

The coating is illuminated with the light of the appropriate energy (color, frequency -f_{ex}) to excite the coating-entrapped probe molecules. The resulting luminescence output (frequency -f_{em}) is inversely proportional to the surface pressure or temperature of the test model.

The PSP/TSP measurement technique is based on the deactivation of photochemically exited molecules, luminophores, and by the presence of oxygen molecules. When luminophores absorb light with the correct wavelength, they are promoted from their base energy state to a higher state. These molecules can lose the extra energy either through emission of light (luminescence) or even without radiation. The interaction of excited luminophores with oxygen molecules, i.e., quenching increases the probability of luminosity, which is recognizable on the surface of the blade. Such a fluorescent image arising under the flow conditions can be recorded using a detector with the same properties as a covering paint or CCD camera with an optical filter. For the purpose of increasing adhesive properties, we added to the composition silicon carbide paint. We also must be confident that this composition will be stable under vibration and shocks.

6.4.3 Application of the PSP/TSP Technique

In an operational test, the wind turbine blade model was coated with oxygen-sensitive probe molecules/polymeric formulation and illuminated with the light of appropriate energy (color). The resulting luminescence from the model is imaged using a charge-couple device (CCD) camera. Pressure is correlated with the ratio of PSP images acquired at a reference condition of known pressure and temperature and at a condition through a modified form of the Stern–Volmer relationship.

$$\frac{Iref}{Ip} = A + B\frac{P}{Pref} \tag{6.35}$$

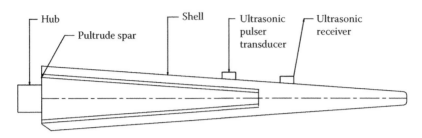

Ultrasonic pulser control

FIGURE 6.9
Ultrasonic pulser control.

In this approach, a reference intensity (Iref) at a given reference pressure (Pref) is divided by the luminescence intensity (Ip) at some test condition (P) over the region of interest. The coefficients A and B are intensity ratio (Iref/Ip), lifetime (τ), and are then correlated with the output of the CCD, providing a convenient tool for the generation of a spatially continuous pressure map, allowing the entire test surface to be sampled simultaneously. Current CCD cameras have a million unequaled spatial resolutions. The output of the CCD array can be visually represented as a two-dimensional image, with the luminescence corresponding to a gray or false-color scale. It is more difficult to use the luminescent light in the water.

Immersion transducers are specifically designed to transmit ultrasound waves in a situation where the test part is partially or wholly immersed in the water. Typically, immersion transducers are used inside the water tank in scanning applications.[30] Ultrasonic pulser controls work automatically if the pulser transducer and ultrasonic receiver are maintained at one frame and drive from one motor.

In Figure 6.9, pos. 1 represents the controlled ultrasonic pulser transducer (for example, model 5900 PR Panametrics Inc., Waltham, MA). Pos. 1 transducers are used for transmitting, and pos. 2 for receiving the ultrasound longitudinal waves. Pos. 3 represents a polymer coating. The sound generated with the frequency range from 20 to 250 kHz and a velocity of ultrasound propagation c could be determined as:

$$c = \lambda/T \text{ or } c = \lambda f \qquad (6.36)$$

where:
λ is the wavelength of one cycle ultrasound waves propagation, m;
T is a period of complete cycle, sec;
f is the frequency of longitudinal wave propagation, 1/sec.
Our concept for experimental pressure determination based on empirical correlation is:

$$f_{em} = f_{ex} \log t \log P \, T \qquad (6.37)$$

where:

f_{em} is the output receiving frequency;

f_{ex} is the transmitting input frequency;

t is a temperature;

P is the pressure under elevation;

T is the period of impulse propagation.

We input T from Equation (6.36) and the result is Equation (6.38):

$$k = \frac{f_{em}}{f_{ex}} * \frac{c}{\log t\lambda} \qquad (6.38)$$

The critical value of the pressure p_{mn} can be determined using Equation (6.39),[13] which covers the conditions for covering stability.

$$P_{mn} = \frac{\pi^2}{h^2} \left(\frac{g}{\lambda h} \right)^{1/2} \left[D_1 \left(\frac{m}{k} \right)^4 + 2D_3 n^2 \left(\frac{m}{k} \right)^2 + D_2 n^4 \right]^{1/2} (1 + \delta) \qquad (6.39)$$

Here:

K = L/h;

δ is the damping coefficient;

D_{ij} is the stiffness of the polymer covered by using PSP's technique.

$$D_1 = Q_{11} \frac{h^3}{12}; \quad D_2 = Q_{22} \frac{h^3}{12}; \quad D_3 = \frac{h^3}{12} (Q_{12} + 2Q_{66}); \qquad (6.40)$$

Stiffness constants Qij are determined from Equation (6.41).

Here:

h is the height of the covering;

λ is the wavelength of the cycle of ultrasound wave propagation ($\lambda = c \, T$);

g is the density of the polymer covering;

η is the acceleration due to gravity;

L is the length of the covering;

m is the digital number of the semiwaves in the x direction;

n is the digital number of the semiwaves in the y direction.

Here: Q_{iknm} are the stiffness constants for components that are satisfied by the symmetrical conditions:

$$Q_{iknm} = Q_{kinm} = Q_{ikmn} = Q_{nmik}$$

$$Q_{1111} = \frac{E_{11}}{1 - \mu_{12}\mu_{21}}; \quad Q_{2222} = \frac{E_{22}}{1 - \mu_{21}\mu_{12}}; \tag{6.41}$$

$$Q_{1122} = \frac{E_{11}\mu_{22}}{1 - \mu_{12}\mu_{21}}; \quad Q_{2211} = \frac{E_{22}\mu_{11}}{1 - \mu_{21}\mu_{12}}; \quad Q_{1212} = G_{12}; \quad Q_{1122} = Q_{2211}$$

Above E_{11}, E_{22}, G_{12}, μ_{11}, μ_{22} are Young's modulus, shear modulus of elasticity, and the Poisson ratio in warp and fill directions.

The last time, vanadium dioxide had been used as infrared active coating.[32]

Thermochromic materials are characterized by a semiconductor-to-metal transition occurring from a reversible change in their crystalline structure as a function of the temperature.

This change has been observed in transition metal oxides,[33,34] such as Ti_2O_3, Fe_3O_4, Mo_9O_{26}, and in several phases of vanadium oxide V_nO_{2n-1}. Among them, VO_2 has received the most attenuation because of the large reversible change of electric, magnetic, and optical properties when temperature was around 70°C.

6.4.4 Luminescent Paint Control

For luminescent paint control, we used a portable and conventional infrared camera. The analyzer can accept interchangeable sampling modules that include a "spotter" module that detects surface luminescence and a bifurcated fiber optic contact probe. It requires no consumables or technical expertise, and performs real-time analysis. Its computerized design measures all the luminescent properties of the compounds within seconds by multidimensional luminescence (MDL).

The MDL information is best displayed three dimensionally, i.e., emission wavelength, excitation wavelength, and wavelength dependent on intensity characteristics. The spectral maps resemble mountainous terrain and are represented in Mott.[33] Description of the infrared cameras is shown in Figure 6.10.

An ultraviolet (UV) lamp is the source represented by the Xenon-flash lamp and flashes at 20 different excitation wavelengths in less than one second, and at each wavelength flash, a diode array detector simultaneously measures all spectral emission characteristics. Most commercial thermo cameras use a high-pressure xenon continuum as a UV light source.

The source illuminates a scanning monochromator entrance slit. Monochromatic light from the exit slit of the excitation monochromator is focused on the sample to be analyzed. If the sample has luminescent properties, it will emit light with specific spectral characteristics.

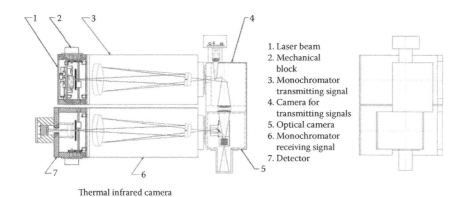

1. Laser beam
2. Mechanical block
3. Monochromator transmitting signal
4. Camera for transmitting signals
5. Optical camera
6. Monochromator receiving signal
7. Detector

Thermal infrared camera

FIGURE 6.10
Thermal infrared camera.

During the molecular excitation process, the sample can receive light of a narrow band of wavelength for quantitative analysis, or the sample can receive light of progressively changing wavelengths (scanning) for qualitative identification of investigative compounds.

The luminescence that is excited in the sample by the excitation monochromator is viewed by the emission monochromator's entrance slit. The emission monochromator spectrally resolves the molecular emission characteristics of the sample and the emitted light intensity at each dispersed wavelength is registered on a photo detector at the exit slit of the emission monochromator. A photo multiplier detector is usually employed as a detector, and the monochromator's dispersing element is rotated by a scanning motor. The scanning operation of the emission monochromator is generated as an emission spectrum.

In a similar fashion, the scanning operation of the excitation monochromator generates an excitation spectrum. The excitation spectrum has the most characteristics, since it generally has more distinctive features and it relates to a compound in normal UV and absorption spectrum. The energy of UV light can be determined by the classic formula:

$$E = h\frac{C}{\lambda} \tag{6.42}$$

Here:
E is the energy in ergs (the erg is a unit of energy in the metric system);
The h is a proportional constant known as "Planck's Constant," which has the value of $6.62*10^{-27}$ erg sec./photon;
C is the speed of light in a vacuum of C, which equals $3.00*10^{10}$ cm/sec.;
λ is the wavelength in cm. For UV light, the spread of 250 to 475 cm is the wavelength;

Frequency is defined as the number of waves or cycles that pass a fixed point in one second.

$$v = \frac{C}{\lambda} \tag{6.43}$$

Energy that is necessary is calculated, if the UV light spread of 250 to 475 cm wavelength is in a vacuum. As noted earlier, the $3.00*10^{10}$ cm/sec. is the speed of light in a vacuum.

When traveling through a material medium, such as fiberglass, the light has a speed slightly different from that in a vacuum. A parameter called the refractive index, N, is defined as:

$$N = \frac{Cv}{Cm} \tag{6.44}$$

Here:
   Cv is the speed in a vacuum;
   Cm is the speed in a medium.

This parameter is easily measured on a device known as a refractometer.

In Figure 6.10, a block is shown for control of optical operations that is used as one principal component for the spectrofluorometer. A stepping motor, pos. 1, rotates antibacklash gears; pos. 3 moves a linear rack; pos. 4 connects with a diffraction grating pos. 7. A beam in pos. 6 creates the Xenon-flash lamp pos. 5 that travels from the slit through the diffraction grating and goes on to the excitation monochromator. Figure 6.10 represents all the thermo camera components beginning with the block operations control pos. 1 to the excitation monochromator pos. 3 through the modules pos. 4, 5 to the emission monochromator pos. 6. Finally the detector is represented by module 7. All luminescence measurements are made on a radiometric basis.

The integrated intensity of luminescence generated by a given flash lamp firing is rationed against the integrated intensity of the flash lamp and flash itself. A small PC board contains a photodiode and associated analog circuitry.

Recently, the FLIR System used infrared detector material, vanadium oxide, which detects vanadium coatings.

The output of the analog radiometer is coupled to the microprocessor system. For UV light stabilization, Hoechst Celanese Corporation (Summit, New Jersey) recommends adding polymers THPE-BZT stabilizer, which consists of 1,1,1-tris (hydroxyphenyl)ethane benzotriazole.

6.4.5   Experimental Investigation

A preliminary estimate on the applications of the different techniques for nonintrusive pressure measurement systems, based on analysis references, have approved the use of the immersion acoustic transducers specifically designed for this purpose.

A wind turbine blade model was designed using PSP-active polymers in a water tunnel/channel. Document calibration and comparable average and time-accurate pressures (kHz) with pressure tap measurement were included in the test was performed any change value of hydrostatic pressure depend of high elevation and frequency vibration represents in Table 6.16.

Suppose that a light beam has a wavelength of 500 nm in free space and then enters a new medium where the speed of light is only 2.00×10^8 m/ sec. (in vacuum 3.00×10^8 m/sec.). Algorithm was correlated wavelength λ and hydrostatic pressure p depends from some parameters illustrated in the Equation (6.45):

$$\lambda = \frac{pRV}{Q_{ij}\delta\alpha_i} \tag{6.45}$$

Here:

p is the hydrostatic pressure;
R is the middle radius of the solid structure;
V is the velocity of light propagation;
Q_{ij} is a stiffness constant;
δ is the middle thickness of structure;
α_l is the coefficient of the linear expansion.

Algorithms were gotten from the condition that is in correlation between strength and stiffness, Equation (6.46)

TABLE 6.16

Change Value of Hydrostatic Pressure Depends on High Elevation and Frequency Vibration

Value of Hydrostatic Pressure, psi/ kg/cm²	Height of Elevation, ft	Frequency of Vibration, Hz	Wavelength, λ, nm	Light Color
14.0/1.0	1.0	1.0	78.3	Blue
58.45/4.13	100.0	100.0	323.0	Blue
102.9/7.26	200.0	200.0	568.0	Green
157.35/10.4	300.0	300.0	813.0	Orange
201.8/13.52	400.0	400.0	1058.0	Black
246.25/16.65	500.0	500.0	1300.0	Black

$$\sigma_{ij} = Q_{ij}(\varepsilon_{ij} - \alpha_{ij}T) \tag{6.46}$$

mechanical deformations equal zero and exist only in the thermal deformations.

The hydrostatic pressure is equal in different directions and distributed as:

$$Q_{ij} = \frac{pR}{\delta} \tag{6.47}$$

We input in Equation (6.46) and Equation (6.47), so that the temperature relative to other parameters has a significance:

$$T = \frac{pR}{Q_{IJ}\delta\alpha_{ij}} \tag{6.48}$$

We calculate λ using parameters: $p = 1$ atmosphere (1 kg/cm^2); $V = 2 \times 10^{10}$ sm/sec; $R = 22.86$ cm (9″); $\delta = .95$ cm; $Q_1 = 3.56 \times 10^5$ kg/sm^2; $\alpha 1 = .0106$cm/cm °C; the result is $\lambda = 78.3$ nm (see Table 6.16).

Whenever a beam crosses a boundary between two layers with molecules/polymeric binder and base coat/primer having different refractive indices, the distribution of power between the transmitted and reflected components is determined by the Fresnel equation:[35]

$$R = \left\{ \frac{n_2 - n_1}{n_2 + n_1} \right\} \tag{6.49}$$

$$T = \frac{4n_2 n_1}{\left(n_2 + n_1\right)^2} \tag{6.50}$$

where:

n_1 and n_2 are the indices on each side of a boundary.

The reflected power per surface of nose cap (R) and the transmitted power per surface of nose cap (T) are used as follows:

$$\phi_{reflected} = \phi_{incident} \times R \tag{6.51}$$

$$\phi_{transmitted} = \phi_{incident} \times T \tag{6.52}$$

where the symbol ϕ represents power (W).

We assumed that the absorption of paint to be negligible. If the block of sensitive paint were immersed in air, where a sunbeam had a refractive index similar to optical glass, R would be 4 percent and the overall transmitted power through the sensitive paint would be 92 percent. However, if the sensitive paint was made of a high-index medium, say n = 3.0, then the power reflection coefficient R would be 25 percent, with an overall transmitted power of 56.25 percent. For such high-index sensitive paint, the use of molecular polymeric binder is necessary to reduce such large reflection losses. Beams reflected from the top of the sensitive paint, thus, will be 180°C out of phase with the beams that have been reflected from the base coat primer substrate interface after having travelled one round trip through the sensitive paint.

6.4.6   Concluding Remarks

1. The power wind turbine blades model with a protective covering of luminescent paint can be used for nonintrusive pressure (PSP) and temperature (TSP) measurements. The nonintrusive pressure/temperature measurement system can be used by nonstandard conditions, such as storms, hurricanes, etc., for determining the turbulent layer parameters.

2. Smart paint material managed from cover layers can protect wind turbine blades from impact loads and reduce noise and vibration.

3. The information about changing hydrostatic pressure, temperature, and moisture can be developed to help aerodynamic characteristics and safe conditions.

4. UV/visible light can help launch a service to regulate turbine blade driving procedures.

References

1. Golfman, Y. 1966. Influence the gauges loops for measurement stress in 0, 45, 90 degree. Paper presented at the Science Conference of Forest Academy, Leningrad, USSR, June.
2. Rose, J. L., 1999, *Ultrasonic waves in solid media*, Cambridge University Press, New York.
3. Rose, J.L. and B.B. Goldberg. 1979. *Basic physics in diagnostic ultrasound*, Wiley & Sons, New York.
4. Golfman, Y. 1993. Ultrasonic non destructive method to determine modulus of elasticity of turbine blades. *SAMPE Journal* 29 (4).
5. Gershberg, M. 1966. Nondestrictive methods for control shipbuilding materials, *Journal of Shipbuilding Technology* 8.
6. Golfman, Y., and M. Gershberg, 1974. Estimate parameters of elasticity using nondestrictive ultrasonic method, *Journal of Shipbuilding Technology* 74.

7. Lekhnitskii, S. G. 1981. Theory of elasticity of anisotropic body. Moscow: Mir Publishers (Trans. in English).

8. Miller, I. K., J. E. Freund, and R .Jonson 1989. *Probability and statistics for engineers*, 4th ed., Englewood Cliffs, NJ: Prentice Hall.

9. Committee of Standard Measures and Devices. 1966. *Methods of statistical treatment of the experimental datum.* Moscow: Committee of Standard Measures and Devices.

10. Golfman, Y. 2001. Nondestructive evaluation of aerospace components using ultrasound and thermography technologies. *Journal of Advanced Materials* 33 (4).

11. Golfman, Y. 1994. Effect of thermoelasticity for composite turbine disk. Paper presented at the 26th International SAMPE Technical Conference, Atlanta, GA,

12. Agfa Nondestructive Testing. 2002. *Agfa integrates NDT technologies. High-performance composites.* Lewistown, PA: Agfa.

13. Golfman, Y. 2003. Dynamic aspects of the lattice structures behavior in the manufacturing of carbon-epoxy composites. *Journal of Advanced Materials* 35 (2).

14. NASA. 2000. *NASA uses radiance infrared camera for component evaluation.* Technical Brief, 24 N.1. Washington, D.C.: NASA.

15. Golfman, Y. 2005. Dynamic local mechanical and thermal strength prediction using NDE for material parameters evaluation of aerospace components. *JAM* 35 (1).

16. Timoshenko, S.P. and J. N. Goodier. 1957. *Theory of elasticity*, McGraw-Hill, New York.

17. Tsai, S. W., and E. M. Wu. 1971. A generalized theory of strength for anisotropic materials. *Journal of Composites Materials* 5.

18. Wu, E. M. 1974. Phenomenological anisotropic failure criteria. In *Mechanics of composite materials*, Vol. 2., ed. G .P. Sendeckyj. San Diego: Academic Press.

19. Hoffman, O. 1967. The brittle strength of orthotropic materials. *Journal of Composite Materials* 1.

20. Hill, R. 1950. *The mathematical theory of plasticity*. Oxford University Press.

21. Golfman, Y. 1991. Strength criteria for anisotropic materials. *Journal of Reinforced Plastics & Composites* 10 (6).

22. Tsai, S. W. 1986. *Composites design.* Dayton, OH: U.S. Air Force Materials Laboratory.

23. Gamstedt, E. K., and R. Talrega. 1999. Fatigue damage mechanisms in unidirectional carbon-fiber-reinforced plastics. *Journal of Material Science* (34): 2535–46.

24. Silva Munoz, R. and A. Lopez Anido. 2008. Monitoring of marine grade composite double plate joints using embedded fiber optic strain. *Journal of Advanced Materials* 4.

25. Golfman, Y. 2004. The fatigue strength prediction for aerospace components using reinforced fiberglass or graphite/epoxy. *Journal of Advanced Materials* 36 (2).

26. Phoenixx TPC. 2004. Thermo-Lite thermoplastic composites, SAMPE Chapter Meeting, Boston Chapter, October 14.

27. Kessler, M. R., N. R. Sottos, and S. R White. 2003. Self-healing structural composite materials. *Composite Part A* 34: 744–753.

28. Golfman, Y. 2010. *Hybrid anisotropic materials for structural aviation parts.* Boca Raton, FL: Taylor & Francis Group.

29. Golfman, Y., E. Ashkenazi, L. Roshkov, and N. Sedorov. 1974. *Machine components of fiberglass for shipbuilding.* Leningrad: Shipbuilding.

30. Gross, L. P., T. W. McGray, D. D. Trump, B. Sarka, and J. C. P. N. Davis. 2001. Pressure-sensitive paint technology at ISSI. (Report) Dayton, OH: Innovative Scientific Solutions Inc. Online at: www.innssi.com

31. Engler, R. H., C. Klein, and O. Thinks. 2000. Pressure sensitive paint measurements at a transonic compressor stage. Paper presented at the Proceedings of the Symposium Measurement Techniques in Transonic Flow, Florence, Italy, September 21–22.

32. Guzman, G. 1968. Vanadium dioxide as infrared active coating. Online at: http://www.solgel.com/articles/August00/thermo/Guzman.htm

33. Mott, N. F. 1968. Non-crystal solids, *Reviews of Modern Physics* 40 (4): 677.

34. Adler, D. 1968. Effect of pressure of the electrical conductivity of sapphire, *Reviews of Modern Physics* 40 (4): 714.

35. Boreman, G. D. 1998. Basic electro-optics for electrical engineers. Bellingham, WA: SPIE Optical Engineering Press.

7

Aerodynamic Structural Noise

7.1 Introduction

Low-noise wind turbine blades are being manufactured with hybrid fiber-glass-reinforced materials that are capable of bearing directional loads and are used in the construction of wind turbine blade parts or entire sections. The use of such materials influences the air flow over them in such a way that the resulting boundary-layer turbulence is damped in a controlled way, thus weakening the noise-scattering mechanism at the trailing edge, and the scattered acoustic waves are absorbed and attenuated by the material acting as an acoustic liner.

7.2 Wind Turbine Aerodynamics

The wind turbine aerodynamics of a horizontal axis wind turbine (HAWT) are not straightforward. The air flow at the blades is not the same as the airflow farther away from the turbine. The very nature of the way in which energy is extracted from the air also causes air to be deflected by the turbine. In addition, the aerodynamics of a wind turbine at the rotor surface exhibit phenomena that are rarely seen in other aerodynamic fields.[1] Wind speed increase during hurricanes is shown in Figure 7.1.

Wind turbines with vibration damping are acoustically optimized to reduce the resonant vibrations and enhance the sound transmission loss of the blade structures. Damping-covering layers also provide maximum absorption of sound with minimum thickness and weight.

7.2.1 Axial Momentum and the Betz Limit

We can show schematically upwind and downwind rotor emissions. Energy in fluid is contained in four different forms: gravitational potential energy, thermodynamic pressure, kinetic energy from the velocity, and, finally,

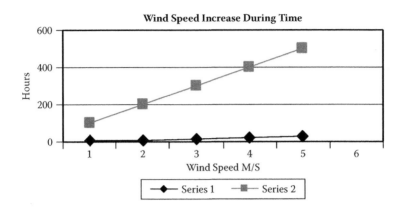

FIGURE 7.1
Wind speed increase during hurricanes is shown. (From University of Massachusetts. 2011. Wind turbine aerodynamics measuring wind turbine noise. Online at www.en.wikipedia/ wiki/Wind_turbine_aerodynamics. With permission.)

thermal energy. Gravitational and thermal energy have a negligible effect on the energy extraction process. From a macroscopic point of view, the air flow about the wind turbine is at atmospheric pressure. If pressure is constant, then only kinetic energy is extracted. However, up close near the rotor itself the air velocity is constant as it passes through the rotor plane. This is because of conservation of mass. The air that passes through the rotor cannot slow down because it needs to stay out of the way of the air behind it. So, at the rotor, the energy is extracted by a pressure drop. The air directly behind the wind turbine is at subatmospheric pressure; the air in front is under greater than atmospheric pressure. It is this high pressure in front of the wind turbine that deflects some of the upstream air around the turbine (Figure 7.2).

Albert Betz and Frederick W. Lanchester were the first to study this phenomenon.[2] Betz notably determined the maximum limit of wind turbine performance. The limit is now referred to as the *Betz limit*. This is derived by looking at the axial momentum of the air passing through the wind turbine. As stated above, some of the air is deflected away from the turbine. This causes the air passing through the rotor plane to have a smaller velocity than the free stream velocity. The ratio of this reduction to that of the air velocity far away from the wind turbine is called the *axial induction factor*. It is defined as:

$$a \equiv \frac{U_1 - U_2}{U_1} \tag{7.1}$$

where:
a is the axial induction factor;
U_1 is the wind speed far away upstream from the rotor;
U_2 is the wind speed at the rotor.

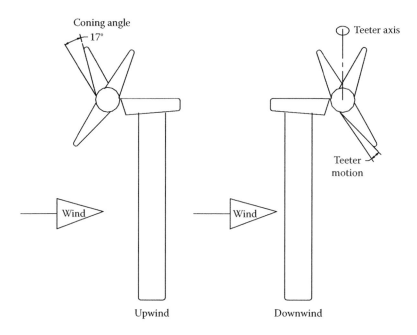

Upwind and downwind rotor emission

FIGURE 7.2
An overview of sound emissions from wind turbines. (From Rogers, A. L. 2006. University of Massachusetts. Online at: www.renewableenergy.world.com. With permission.)

The first step in deriving the Betz limit is to apply conservation of axial momentum. As stated above, the wind loses speed around the wind turbine compared to the speed far away from the turbine. This would violate the conservation of momentum if the wind turbine was not applying a thrust force on the flow. This thrust force manifests itself through the pressure drop across the rotor. The front operates at high pressure while the back operates at low pressure. The pressure difference from the front to back causes the thrust force. The momentum lost in the turbine is balanced by the thrust force. Another equation is needed to relate the pressure difference to the velocity of the flow near the turbine. Here, the Bernoulli equation is used between the field flow and the flow near the wind turbine. There is one limitation to the Bernoulli equation: The equation cannot be applied to fluid passing through the wind turbine; instead, conservation of mass is used to relate the incoming air to the outlet air. Betz used these equations and managed to solve the velocities of the flow in the far wake and near the wind turbine in terms of the far field flow and the axial induction factor. The velocities are given as:

$$U_2 = U_1(1 - a) \tag{7.2}$$

$$U_4 = U_1(1 - 2a) \tag{7.3}$$

U_4 is introduced here as the wind velocity in the far wake. This is important because the power extracted from the turbine is defined by the following equations. However, the Betz limit is given in terms of the coefficient of power. The coefficient of power is similar to efficiency, but not the same. The formula for the coefficient of power is given beneath the formula for power:

$$P = 0.5\rho A U_2 \left(U_1^2 - U_4^2 \right) \tag{7.4}$$

$$C_P \equiv \frac{P}{0.5\rho A U_1^3} \tag{7.5}$$

Betz was able to develop an expression for Cp in terms of the induction factors. This is done by the velocity relations being substituted into power and power is substituted into the coefficient of power definition. The relationship Betz developed is:

$$C_p = 4a(1 - a)^2 \tag{7.6}$$

The Betz limit is defined by the maximum value that can be given by the above formula.

This is found by taking the derivative with respect to the axial induction factor, setting it to zero, and solving for the axial induction factor. Betz was able to show that the optimum axial induction factor is one-third. The optimum axial induction factor was then used to find the maximum coefficient of power. This maximum coefficient is the Betz limit. Betz was able to show that the maximum coefficient of power of a wind turbine is 16/27. Airflow operating at higher thrust will cause the axial induction factor to rise above the optimum value. Higher thrust causes more air to be deflected away from the turbine. When the axial induction factor falls below the optimum value, the wind turbine is not extracting all the energy it can. This reduces pressure around the turbine and allows more air to pass through the turbine, but not enough to account for the lack of energy being extracted.

The derivation of the Betz limit shows a simple analysis of wind turbine aerodynamics. In reality, there is a lot more. A more rigorous analysis would include wake rotation, the effect of variable geometry. The effect of air foils on the flow is a major component of wind turbine aerodynamics. Within airfoils alone, the wind turbine aerodynamicist has to consider the effect of surface roughness, dynamic stall tip losses, solidity, among other problems.

7.3 Measuring Wind Turbine Noise

Measuring decibel levels is extremely important. Decibels are the least important aspect of the various sounds that wind turbine blades make.[3] The decibel (dB) is a logarithmic unit that indicates the ratio of a physical quantity (usually power or intensity) relative to a specified or implied reference level. A ratio in decibels is 10 times the logarithm to base 10 of the ratio of two power quantities.[4] Being a ratio of two measurements of a physical quantity in the same units, it is a dimensionless unit. A decibel is one-tenth of a bel, a seldom used unit.

Power is the amount of energy used (or generated) per unit of time. Sound pressure and sound intensity is shown in Figure 7.3. Sound pressure change with different frequencies is demonstrated in Figure 7.4, and Figure 7.5 illustrates the measuring of sound.

A, B, C—weighting compensates for the sensitivity of the human ear. A, B, C—weighted levels designated as dB(A). Relative response has less level between 63–250 Hz. Sound emissions reduce when frequencies increase (Figure 7.6).

7.4 Reduce Noise in Wind Turbine Blades

Of growing significance is the aerodynamic part of wind turbine noise: CTQ (critical to quality). It is gaining importance from a competitive and

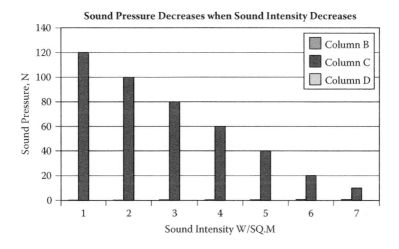

FIGURE 7.3

Sound pressure and sound intensity. (From Rogers, A. L. 2006. University of Massachusetts. Online at: www.renewableenenergy.world.com. With permission.)

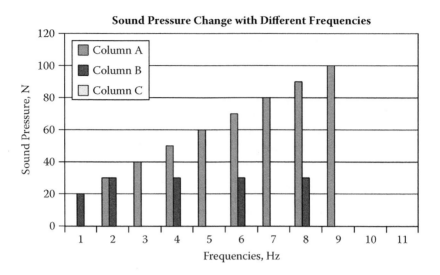

FIGURE 7.4
Sound pressure change with different frequencies. (From Rogers, A. L. 2006. University of Massachusetts. Online at: www.renewableenenergy.world.com. With permission.)

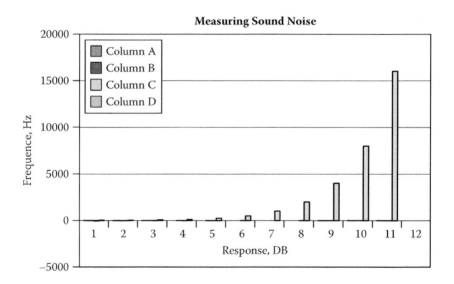

FIGURE 7.5
Measuring sound. (From Rogers, A. L. 2006. University of Massachusetts. Online at: www. renewableenenergy.world.com. With permission.)

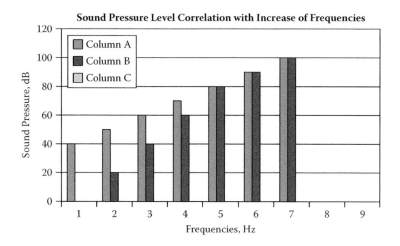

FIGURE 7.6

Sound emissions reduce when frequencies increase. (From Wind turbine acoustic noise. 2006. Online at: www.ceere.org. With permission.)

regulatory perspective as the market is driving product design to larger turbines with higher blade tip speed. Here, aerodynamic noise becomes a crucial constraint on efficiency and yield of a design. As such, there is a need for concepts for noise reduction.

Gupta and Maeder's[3] invention defines a concept wherein materials technology is used as a means of noise reduction, targeting blade self-noise and tip-noise, the two primary components of wind turbine aerodynamic noise. In the past, aerodynamic shaping has been the primary means of achieving lower noise levels, e.g., the use of a large chord, higher solidity blades, low-noise airfoil design, plan-form, and tip/winglets geometry. Noise reduction concepts that use add-ons like trailing edge serrations, sharp trailing edge inserts, and the like, have also been investigated and, in some cases, put into production.

The invention proposes to incorporate composite material in wind turbine blades to reduce noise, via noise source reduction and/or noise attenuation and absorption.

The invention may be embodied in a wind turbine blade having a pressure side and a suction side, a leading edge, a trailing edge, and a tip region, at least a portion of the blade being formed from a cellular material, and the cellular material portion defining a portion of an exposed surface of the blade, whereby aerodynamic noise is reduced via noise source reduction and/or noise attenuation and absorption by the cellular material.

The invention also may be embodied in a method of reducing noise in a wind turbine via at least one of a noise source reduction and noise attenuation and absorption by a wind turbine blade; the method comprising: providing a wind turbine blade for the wind turbine, the blade having a pressure side and a suction side, a leading edge, a trailing edge, and a tip region; wherein at least a portion of the blade is formed from a cellular material and where the cellular material portion defines a portion of an exposed surface of the blade.

Different wind velocities in the top-Vp and bottom-Vb points of blades at an elevation of approximately 200 m creates turbulence and vibration.

The ratio of reduction air velocity is represented in equation (7.7).

$$\alpha_1 = \frac{Vt - Vb}{Vb} \tag{7.7}$$

and power is given by:

$$P = 0.5\rho A \, V_{mid} \, (V - Vp) \tag{7.8}$$

Here:
$$V_{mid} = \frac{Vt - Vb}{2} \tag{7.9}$$

This phenomenon will be increased noise (Figure 7.7).

For reduced noise between flange fingers bush and hub, we install shock absorbers (see Figure 7.7). Shock absorber pads dampen vibrations and reduce noise. Shock absorbers are seen in Figure 7.8.[5]

Another solution can be plate springs or nylon bushes. But, the advantage of shock absorbers is the *automatic* regulation of vibration and noise. Pneumatic

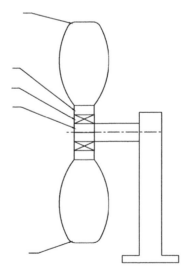

Wind turbine blade, aerodynamic aspect

FIGURE 7.7
Assembly blades with shock absorber.

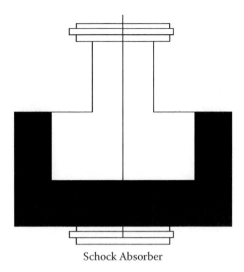

Schock Absorber

FIGURE 7.8
Shown is a shock absorber to dampen vibration and noise.

and hydraulic shock absorbers commonly take the form of a cylinder with a sliding piston inside. The cylinder is filled with a fluid, such as hydraulic fluid, or air. This fluid-filled piston/cylinder combination is a dashpot. Shock absorbers include cushions and springs. The shock absorber's duty is to absorb or dissipate energy. One design consideration, when designing or choosing a shock absorber, is where that energy will go. In most dashpots, energy is converted to heat inside the viscous fluid.

7.5 Sound Emissions, Temperature and Pressure

Upwind rotor emissions and downwind rotor emmisions are shown in Figure 7.2. Sound and pressure wind waves distribution are represented in Figure 7.4.

Sound has different frequencies, e.g., human hearing: 20 to 20,000 Hz, infrasound less than 20 Hz. The threshold level of hearing at 10 Hz is equal to 100 dB. Following Rogers,[2] the standard deviation of threshold of perception level is 6 dB.

Broadband sound power level L_{wa} versus wind speed correlates with normal distribution law. Noise propagation model works as:

$$L_p = L_w - 10 \log_{10}(2\pi N^2) - \alpha N \tag{7.10}$$

Here:

L_w is a sound power level equal to 102;

N is the distance from the turbine tower, in meters;

α is a sound absorption coefficient $\alpha = 0.005$ dB/m

7.5.1   Density, Temperature, and Pressure Correlation

The density of dry air can be calculated using the ideal gas law, expressed as a function of temperature and pressure:

$$\rho = \frac{p}{R \cdot T} \tag{7.11}$$

where:

ρ is the air density;

p is absolute pressure;

R is the specific gas constant for dry air;

T is absolute temperature.

The specific gas constant for dry air is 287.058 J/(kg·K) in SI units, and 53.35 (ft·lbf)/(lbm·R) in U.S. customary and Imperial units.[6]

Therefore:

- At IUPAC standard temperature and pressure (0°C and 100 kPa), dry air has a density of 1.2754 kg/m³.
- At 20°C and 101.325 kPa, dry air has a density of 1.2041 kg/m³.
- At 70°F and 14.696 psia, dry air has a density of 0.074887 lbm/ft3.

Table 7.1 illustrates the air density–temperature relationship at 1 atm or 101.325 kPa:

7.5.2   Water Vapor

The addition of water vapor to air (making the air humid) reduces the density of the air, which may at first appear contrary to logic.[7]

This occurs because the molecular mass of water (18 g/mol) is less than the molecular mass of dry air (around 29 g/mol). For any gas, at a given temperature and pressure, the number of molecules present is constant for a particular volume (see Avogadro's law). So, when water molecules (vapor) are added to a given volume of air, the dry air molecules must decrease by the same number to keep the pressure or temperature from increasing. Hence, the mass per unit volume of the gas (its density) decreases.

The density of humid air may be calculated as a mixture of ideal gases. In this case, the partial pressure of water vapor is known as the vapor pressure.

TABLE 7.1

Air Density–Temperature Relationship

	Effect of Temperature		
Temperature ϑ in °C	Speed of Sound c in m·s−1	Density of Air ρ in kg·m^{-3}	Acoustic Impedance Z in N·s·m^{-3}
+35	351.96	1.1455	403.2
+30	349.08	1.1644	406.5
+25	346.18	1.1839	409.4
+20	343.26	1.2041	413.3
+15	340.31	1.2250	416.9
+10	337.33	1.2466	420.5
+5	334.33	1.2690	424.3
±0	331.30	1.2920	428.0
-5	328.24	1.3163	432.1
-10	325.16	1.3413	436.1
-15	322.04	1.3673	440.3
-20	318.89	1.3943	444.6
-25	315.72	1.4224	449.1

Using this method, error in the density calculation is less than 0.2 percent in the range of −10°C to:

$$\rho \text{ humid air} = \frac{p_d}{R_d \cdot T} + \frac{p_v}{R_v \cdot T} \tag{7.12}$$

where: 50°C. The density of humid air is found by:
ρ = Density of the humid air (kg/m³);
p_d = Partial pressure of dry air (Pa);
R_d = Specific gas constant for dry air, 287.058 J/(kg·K);
T = Temperature (K);
p_v = Pressure of water vapor (Pa);
R_v = Specific gas constant for water vapor, 461.495 J/(kg·K).

The vapor pressure of water may be calculated from the saturation vapor pressure and relative humidity. It is found by:

$$p_v = \phi \cdot p_{\text{sat}} \tag{7.13}$$

where:
p_v = Vapor pressure of water;
ϕ = Relative humidity;

p_{sat} = Saturation vapor pressure.

The saturation vapor pressure of water at any given temperature is the vapor pressure when relative humidity is 100 percent. A simplification of the regression[2] used to find this can be formulated as:

$$p(mb)_{sat} = 6.1078 \cdot 10^{\frac{7.5T - 2048.625}{T - 35.85}} \qquad (7.14)$$

NOTE:

- This will give a result in mbar (millibar), 1 mbar = 0.001 bar = 0.1 kPa = 100 Pa.
- p_d is found considering partial pressure, resulting in:

$$p_d = p - p_v \qquad (7.15)$$

where p simply notes the absolute pressure in the observed system.

Correlation vapor pressure with temperature (Figure 7.9). Standard atmosphere: $p_0 = 101325$ Pa, $T_0 = 288.15$ K, $\rho_0 = 1.225$ kg/m³. To calculate the density of air as a function of altitude, one requires additional parameters. They are listed below, along with their values according to the International Standard atmosphere, using the universal gas constant instead of the specific one:

- Sea level standard atmospheric pressure $p_0 = 101325$ Pa

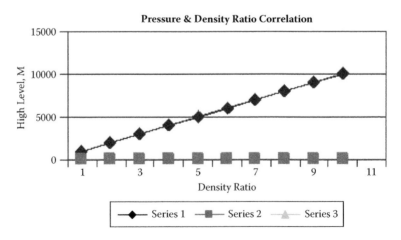

FIGURE 7.9
Pressure and density correlation.

- Sea level standard temperature $T_0 = 288.15$ K
- Earth surface gravitational acceleration $g = 9.80665$ m/s^2
- Temperature lapse rate $L = 0.0065$ K/m
- Universal gas constant $R = 8.31447$ J/(mol·K)
- Molar mass of dry air $M = 0.0289644$ kg/mol

Temperature at altitude h meters above sea level is given by the following formula (only valid inside the troposphere):

$$T = T_0 - L \cdot h \tag{7.16}$$

The pressure at altitude h is given by:

$$p = p_0 \cdot \left(1 - \frac{L \cdot h}{T_0}\right)^{\frac{g \cdot M}{R - L}} \tag{7.17}$$

Density can then be calculated according to a molar form of the original formula:

$$\rho = \frac{p \cdot M}{R \cdot T} \tag{7.18}$$

7.6 Offshore Support Structures for Power Wind Turbine Blades

7.6.1 Introduction

Support structures for offshore wind turbine blades have a highly dynamic behavior. Complex wind and hydrodynamic loading and permanently changing severe conditions are complicating these tasks. The integrated effect of the wind and wave loads requires flexibility from the wind turbine control systems.

In a severe situation, the total loading is likely to be significant compared to the sum of the constituent loads.

7.6.1.1 Support Structure Design

Support structure design has some requirements:

1. The loads are not coincident
2. Aerodynamic damping is provided by the rotor, which significantly dampens the motions due to wave loading
3. Buckling analysis support structures

Structures to support wind turbines comes in various shapes and sizes (Table 7.2).

The tripod is modeled with a statically determined structure to obtain the stresses in the members. In the model, hinges replace the rigid connections.[9] Simplification of the design process in the different structures (Figure 7.10 to Figure 7.13) can be verified with results. All structures assumed that Euler buckling during pile driving is not a design driver. However, the piles of the guys (cable guides) are slender and deep and a buckling analysis may be required.[10]

7.6.2 Conclusion

1. Jacket structures become more deeply attractive in water.
2. Monopole 2 Tripod structures made from steel tube have very good stability.
3. Gravity 2 floating structures are still under development.

7.6.3 Offshore Wind Initiative

(The following is a news release in January 2011 from the U.S. Departments of Energy and the Interior concerning major offshore wind intiatives.[11] The strategic plan [$50 million in R&D funding] identifies Wind Energy Areas that will speed offshore wind energy development.)

> NORFOLK, VA — Unveiling a coordinated strategic plan to accelerate the development of offshore wind energy, U.S. Secretary of the Interior Ken Salazar and Secretary of Energy Steven Chu today announced major steps forward in support of offshore wind energy in the United States, including new funding opportunities for up to $50.5 million for projects that support offshore wind energy deployment and several high-priority Wind Energy Areas in the mid-Atlantic that will spur rapid, responsible development of this abundant renewable resource.
>
> Deployment of clean, renewable offshore wind energy will help meet the president's goal of generating 80 percent of the nation's electricity from clean energy sources by 2035.
>
> "The mid-Atlantic Wind Energy Areas are a key part of our 'Smart from the Start' program for expediting appropriate commercial-scale wind energy development in America's waters," Secretary Salazar said. "Through the Strategic Work Plan, the United States is synchronizing new research and development initiatives with more efficient,

TABLE 7.2

Support Structure Options

Structure	Examples	Use	Notes
• Monopile	• Utgrunden (SE), Blyth (UK), Horns Rev (DK), North Hoyle (UK), Scroby Sands (UK), Arklow (IE) Ireland, Barrow (UK), Kentish Flats (UK), OWEZ (NL), Princess Amalia (NL)	• Shallow to medium water depths	• Made from steel tube, typically 4–6 m in diameter • Installed using driving and/or drilling method. Transition piece grouted onto top of pile
• Jacket	• Beatrice (UK), Alpha Ventus (DE)	• Medium to deep water depths	• Made from steel tubes welded together, typically 0.5–1.5 m in diameter • Anchored by driven or drilled piles, typically 0.8–2.5 m in diameter
• Tripod	• Alpha Ventus (DE)	• Medium to deep water depths	• Made from steel tubes welded together, typically 1.0–5.0 m in diameter • Transition piece incorporated onto center column • Anchored by driven or drilled piles, typically 0.8–2.5 m in diameter
• Gravity base	• Vindeby (DK), Tuno Knob (DK), Middlegrunden (DK), Nysted (DK,) Lilgrund (SE), Thornton Bank (BE)	• Shallow to medium water depths	• Made from steel or concrete • Relies on weight of structure to resist overturning, extra weight can be added in the form of ballast in the base • Seabed may need some careful preparation • Susceptible to scouring and undermining due to size
• Floating structures	• Karm øy (NO)	• Deep to very deep water depths	• Still under development • Relies on buoyancy of structure to resist overturning • Motion of floating structure could add further dynamic loads to structure • Not affected by seabed conditions

Source: Zaaijer, M. B. 2001. Offshore support structures for large-scale wind turbines. Online at: www.ecn.nl/docs/dowec/2001-EWEA-STC-Support-structures.pdf. With permission.)

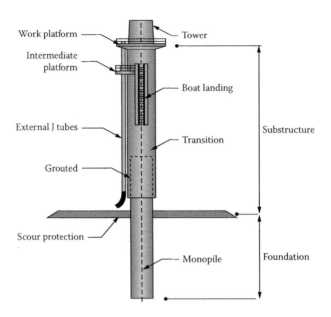

FIGURE 7.10
Monopole structure. (From Zaaijer, M. B. 2001. Offshore support structures for large-scale wind turbines. Online at: www.ecn.nl/docs/d owec/2001-EWEA-STC-Support-structures.pdf. With permission.)

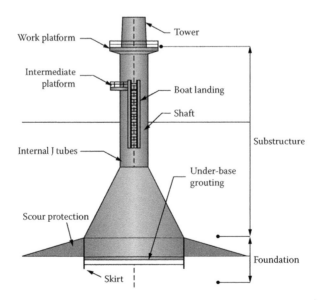

FIGURE 7.11
Graving base structure. (From Zaaijer, M. B. 2001. Offshore support structures for large-scale wind turbines. Online at: www.ecn.nl/docs/dowe c/2001-EWEA-STC-Support-s tructures.pdf. With permission.)

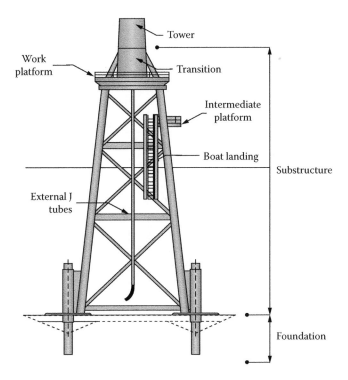

FIGURE 7.12
Jacket structure. (From Zaaijer, M. B. 2001. Offshore support structures for large-scale wind turbines. Online at: www.ecn.nl/docs/dowe c/2001-EWEA-STC-Support-structures.pdf. With permission.)

forward-thinking planning so that we can quickly help start up an American offshore wind industry. This initiative will spur the type of innovation that will help us create new jobs, build a clean energy future, and compete and win in the technologies of the twenty-first century."

"Offshore wind energy can reduce greenhouse gas emissions, diversify our energy supply, and stimulate economic revitalization," said Secretary Chu. "The Department of Energy is committed to working with our federal partners to provide national leadership in accelerating offshore wind energy deployment."

The joint plan, A National Offshore Wind Strategy: Creating an Offshore Wind Industry in the United States, made public today is the first-ever interagency plan on offshore wind energy and demonstrates a strong federal family commitment to expeditiously develop a sustainable, world-class offshore wind industry in a way that reduces conflict with other ocean uses and protects resources.

The plan focuses on overcoming three key challenges: the relatively high cost of offshore wind energy; technical challenges surrounding installation, operations, and grid interconnection; and the lack of site data and experience with project permitting processes.

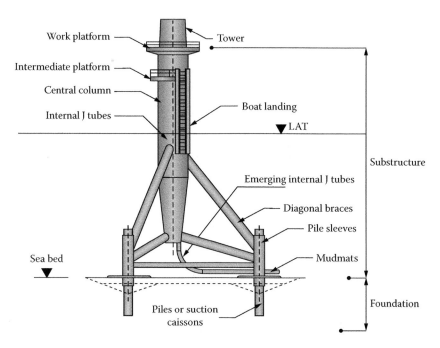

FIGURE 7.13
Tripod structure. (From Zaaijer, M. B. 2001. Offshore support structures for large-scale wind turbines. Online at: www.ecn.nl/docs/dow ec/2001-EWEA-STC-Support-structures.pdf. With permission.)

In support of this Strategic Work Plan, Secretary Chu announced today the release of three solicitations, representing up to $50.5 million over five years, to develop breakthrough offshore wind energy technology and to reduce specific market barriers to its deployment:

Technology Development (up to $25 million over five years): DOE will support the development of innovative wind turbine design tools and hardware to provide the foundation for a cost-competitive and world-class offshore wind industry in the United States. Specific activities will include the development of open-source computational tools, system-optimized offshore wind plant concept studies, and coupled turbine rotor and control systems to optimize next-generation offshore wind systems.

Removing Market Barriers (up to $18 million over three years): DOE will support baseline studies and targeted environmental research to characterize key industry sectors and factors limiting the deployment of offshore wind. Specific activities will include offshore wind market and economic analysis, environmental risk reduction, manufacturing and supply chain development, transmission planning and interconnection strategies, optimized infrastructure and operations, and wind resource characterization.

Next-Generation Drive Train (up to $7.5 million over three years): DOE will fund the development and refinement of next-generation designs for wind turbine drive trains, a core technology required for cost-effective offshore wind power.

Today Salazar also identified four Wind Energy Areas offshore the mid-Atlantic as part of Interior's 'Smart from the Start' approach announced in November 2010 that uses appropriate designated areas, coordinated environmental studies, large-scale planning and expedited approval processes to speed offshore wind energy development.

The areas, on the Outer Continental Shelf offshore Delaware (122 square nautical miles), Maryland (207), New Jersey (417), and Virginia (165), will receive early environmental reviews that will help to lessen the time required for review, leasing, and approval of offshore wind turbine facilities.

In March, Interior also expects to identify Wind Energy Areas off of North Atlantic states, including Massachusetts and Rhode Island, and launch additional NEPA (National Environmental Policy Act) environmental reviews for those areas. A similar process will occur for the South Atlantic region, namely North Carolina, this spring.

Based on stakeholder and public participation, Interior's Bureau of Ocean Energy Management, Regulation and Enforcement (BOEMRE) will prepare regional environmental assessments for Wind Energy Areas to evaluate the effects of leasing and site assessment activities on leased areas.

If no significant impacts are identified, BOEMRE could offer leases in these Mid-Atlantic areas as early as the end of 2011 or early 2012. Comprehensive site-specific NEPA review will still need to be conducted for the construction of any individual wind power facility, and BOEMRE will work directly with project managers to ensure that those reviews take place on aggressive schedules.

Under the National Offshore Wind Strategy, the Department of Energy is pursuing a scenario that includes deploying 10 gigawatts of offshore wind generating capacity by 2020 and 54 gigawatts by 2030. Those scenarios include development in both federal and state offshore areas, including along the Atlantic, Pacific, and Gulf coasts as well as in Great Lakes and Hawaiian waters. Those levels of development would produce enough energy to power 2.8 million and 15.2 million average American homes, respectively.

Today's announcement is the latest in a series of Administration actions to speed renewable energy development offshore by improving coordination with state, local, and federal partners, developing wind research and test facilities for new technologies to reduce market barriers, identifying priority areas for potential development, and conducting early environmental reviews.

References

1. Measuring Wind Turbine Noise. 2010. Online at: www.renewableenergyworld. com. November 22.

2. Rogers, A. L. 2006, Wind turbine noise, infrasound and noise perception. Amherst, MA: University of Massachusetts. Online at: www.ceere.org/rerl

3. Gupta, A. and T. Maeder. 2008. Wind turbine blade and method for reducing noise in wind turbine. US Patent application 20080286110, November 20, 2008. Online at: http://www.faqs.org/patents/ app/20080286110#ixzz1AOMg5wuC

4. Decibel Free Encyclopedia. 2011. Online at: www.en.wikipedia.org/wiki/ Decibel. October 11.

5. Shock absorber. 2000. Wikipedia, the free encyclopedia. Online at: www. en.wikipedia.org/wiki/Shock_absorber

6. International Standard Atmosphere. 2011. Wikipedia. Online at: www. en.wikipedia.org/wiki/International_Standard_Atmosphere, October 11.

7. Water vapor. Wikipedia. 2011. Online at: www.en.wikipedia.org/wiki/Water_ vapor. October 11.

8. Wind Energy–The Facts. 2010. Offshore support structures. Online at: www. wind-energy-the-facts.org/offshore.wind

9. Zaaijer, M. B. 2001. Properties of offshore support structures for large-scale wind turbines. www.tudelft.nl/live/binaries

10. Randolph, M. F. 1981. The response of flexible piles to lateral loading. *Geotechnique* 31 (2).

11. DOE/DOI. 2011. *Offshore wind initiative.* News release. Washington, D.C.: Departments of Energy and the Interior, January 7.

Index